AF517756

NON-CONVENTIONAL LEGUMES AND USEFUL PLANTS OF NORTH EAST INDIA

SUDHIR KUMAR

SCIENTIFIC PUBLISHERS (INDIA)

P.O. BOX 91 JODHPUR

Published by:
Pawan Kumar
Scientific Publishers (India)
5-A, New Pali Road,
P.O. Box 91
Jodhpur - 342 001
E-mail: info@scientificpub.com
www.scientificpub.com

Dr. Sudhir Kumar

© Kumar, Sudhir, 2008

ISBN: 978-81-7233-551-9

Cover design: Tanay Sharma
Lasertype set: Rajesh Ojha
Printed in India

Dedication

The book is dedicated to
my wife Jagriti and
my children Kshamata, Shuchita
and Subodh
and
my student late Dr. Ramesh Sood

PREAMBLE

The North Eastern Region is a rich respositary of useful plants. These plants have not been studied in detail regarding their characteristics, use, utilization, potential and also as a source of valuable protein. This publication is an humber attempt in the above direction and may from the basis of further detailed work. The author is fully conscious of the shortcomings of the present publication. Much more could have been done but because of the problems of insurgency, difficult terrain and lack of literature and information on the plants of the region were the constraints and this is all what the author could do. The author will feel satisfied and amply rewarded if this publication stimulates further efforts and studies on these valuable plants and those listed in the second chapter as compendium of useful plants of the region.

A chapter on useful plants of the region has also been added, so that, all those who may be interested in knowing the plant wealth of this remote region. This part is especially for those, who may be interested in the development and progress of the region and to remove regional imbalance and may be useful for all those, who may be interested in pursuing studies on speicific groups of plants.

Sudhir Kumar

ACKNOWLEDGEMENT

The author is grateful to the Department of Science and Technology (Human Resource Section for retired Scientist) Government of India, New Delhi and Dr. Parveen Farooqui, Advisor and Head, Human Resource Division for personal interest and guidance during preparation and financial support from DST and Prof. H.N. Verma, Head, Department of Botany for writing the Foreword.

CONTENTS

PART - I

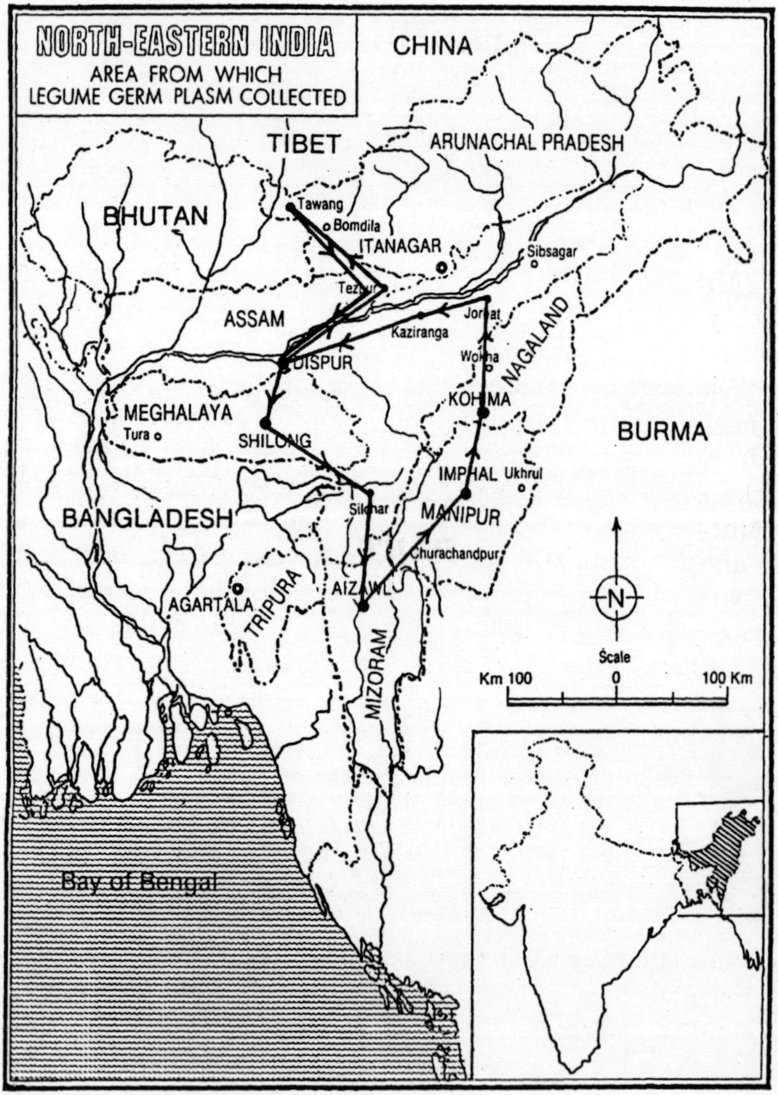
NORTH-EASTERN INDIA
AREA FROM WHICH
LEGUME GERM PLASM COLLECTED
CHINA
TIBET
ARUNACHAL PRADESH
BHUTAN
Tawang
Bomdila
ITANAGAR
Sibsagar
Tezpur
ASSAM
Jorhat
Kaziranga
NAGALAND
Wokha
DISPUR
KOHIMA
MEGHALAYA
Tura
SHILONG
BURMA
IMPHAL
Ukhrul
Silchar
MANIPUR
BANGLADESH
Churachandpur
AGARTALA
TRIPURA
AIZAWL
MIZORAM
N
Scale
Km 100
0
100 Km
Bay of Bengal

PART - I

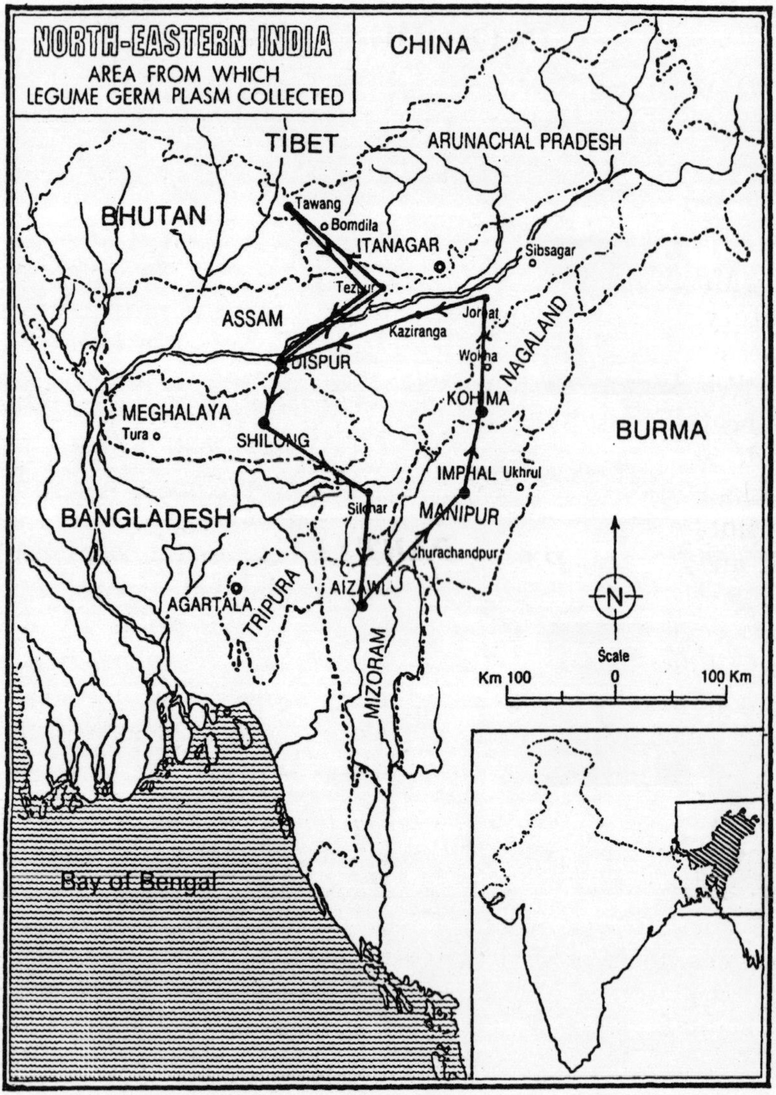
NORTH-EASTERN INDIA
AREA FROM WHICH
LEGUME GERM PLASM COLLECTED
CHINA
TIBET
ARUNACHAL PRADESH
BHUTAN
Tawang
Bomdila
ITANAGAR
Sibsagar
Tezpur
ASSAM
Jorhat
Kaziranga
NAGALAND
Wokha
DISPUR
KOHIMA
MEGHALAYA
Tura
SHILONG
BURMA
IMPHAL
Ukhrul
Silchar
MANIPUR
BANGLADESH
Churachandpur
AGARTALA
TRIPURA
AIZAWL
MIZORAM
Scale
Km 100
0
100 Km
Bay of Bengal

INTRODUCTION

Mankind enters the next millennium. The twentieth century has passed in the hands of the historians for their assessment. We also complete more than half a century of our freedom and establishment of democratic republic. Have we been able to accomplish the pious resolutions which we made when we adopted our constitution. The hunger still starks on the face of millions of our people, inspite of green revolution and bumper crops we had. Our granneries are full and we are even in a position to export but our people still suffer from malnutrition and protein deficiencies as our pulse production is not adequate due to poor yields. By 2040 we will have to feed two billion people and therefore fight on the food front is an unending one. Land being limited and yields reaching a plateau the search for new sources of food will have to be constantly made.

The North Eastern Region, which is still unexplored abounds in large number of leguminous plants which are not only new sources of food but can also be used to supplement the protein deficient diets. The National Science Academy of United States constituted a panel in 1978 t0 identify legumes which could improve the well being of the people in developing countries. This report lists about 200 species of little known legumes of which many are found in North East Region and which have adopted to the warm climate of the region.

The world population now stands at six billion. The increase is more noticeable in the developing countries whose staple food is rice and pulses. The pulses supply the required protein as people of developing countries can ill afford animal protein. The pulses occupy a unique position in Indian agriculture as they provide the only protein component of average Indian diet. The legumes also improve the fertility of the soil and therefore the Indian farmer grows in mixed cropping or grows legume alone to maintain the fertility of the soil as he can not afford costly fertilizer. He also very well knows that use of fertilizer shall deplect the organic matter of the soil resulting in low fertility as well as quality of the soil. India grows maximum number of pulses in the world and has maximum area under pulses in the world (Table 1).

Table 1. Details of area (ha) production (mt) and average yield (Kg per ha) of various pulses grown in India

Name of the pulses	Area		Production		Average yield (kg)	
	2003	2004	2003	2004	2003	2004
Pigeon pea	3,200,000	3,300,000	2,210,000	2,430,000	6,906	7,346
Lentils	1,343,900	1,450,000	833,400	1,100,000	6,201	7,586
Chickpea	5,670,800	6,500,000	4,130,000	5,770,000	7,283	8,877
Peas	730,000	750,000	730,000	800,000	10,000	10,667
Soyabeans	6,450,000	7,550,000	6,800,000	7,550,000	10,543	9,272

Out of the area under pulses Bengal gram represents the largest single item, the Redgram or Arhar comes next. It will be observed that area under pulses has increased from 5% and their contribution as percentage to the total food grain from 12% to 8%. This is real cause for anxiety as the pulses are most acceptable and important part of Indian diet. The practice of taking them with wheat and rice has been found to have a sound scientific basis because these of being rich in lysine supplement the cereals which are deficient in it. It has also been found that taking them with cereals increases the absorbility of pulse protein and does not cause flatulence the anxiety is further aggravated as in early 2015 A.D. about a billion people will have to be fed, therefore, there will be an acute protein shortage. Animal protein is costly and majority of the Indian population being dependent entirely on legumes for supply of proteins in the diet. A massive increase in vegetable protein supply in India presents a less difficult, less expensive and more energy efficient prospect than boosting the supply of animal protein. Daily per capita requirement is of 80 grams which is minimum recommended by WHO. In fact protein availability dropped to 64 grams recently. In 1988-89 it was only 28 grams. Therefore, a revolution similar to the one in cereals can provide the protein requirement needed by Indian population and therefore the battle on the food front has to be relentlessly fought, The reduction in production of grain legumes is highly undesirable and a cause of concern because :

A. The grain legumes are the main source of protein for strict Indian lacto-vegetarian.

B. The legumes maintain and restore the fertility of Indian soils in absence of adequate manuring. The pulses therefore are generally included in crop rotation to keep the soil productive.

C. The pulses have low water requirement and so they play a most important role in dry land cultivation. The dry land constitute 75% of the total cropped area in the country.

Further yields of pulses as compared to cereals are very low. Some of the reasons for low yield are as pulse varieties take a long time to mature, consequently per day production is low. The pulses are grown in poor and marginal lands without the benefit of adequate inputs of good seeds, fertilizer and pesticides. Low level management which has not taken advantage of maximum genetic potential of improved varieties.

The remedial measures therefore require our immediate attention. In addition to correcting causes of low yield it is necessary to search for other leguminous species which may supply badly needed protein as even the total pulse production can not meet the protein requirement of the country.

There are large number of leguminous species which can add to the protein pool of the country. These may be treated or called *Non-conventional Legumes* to differentiate them from traditional pulses which are cultivated on large scale. These non-conventional legumes in addition to supply of proteins have also other diverse economic value. Some of these have medicinal properties due to presence of alkaloids and haemagglutinins whereas others are used as fodders, fibre and dyes. It has also been reported that some of them have toxins which can be used against rats. Some others inactivate human tumour cells as well as have beneficial effects on plastic anaemia. Non-Conventional legumes, therefore, are an important source of much needed proteins in addition to other uses mentioned above.

These vast treasures of legumes grow abundantly in North East Region. Some of them are being invariably used by the majority of people in their diets while most of them still remain neglected. The reasons for their neglect are bitterness of taste, unpleasant flavor, toxicity, ill effects and ignorance. The present food scarcity and protein malnutrition have drawn a good deal of attention towards these legumes. Many of these legumes are used by tribals barring say one third.

Many of these may also be wild progenators of cultivated pulses. Modern breeding is in urgent need of gene pools from which new genes can be introduced into existing cultigens in order to improve their stability and therefore, the importance of these wild legumes increases further because of the possibility of their carrying genes for resistance to insect pests and also for survival in adverse conditions such as drought and cold etc. This naturally necessitates searching for new sources of favorable genes in wild relatives of crop species and preserving the genetic wealth from known centers of biodiversity which are at the moment in jeopardy. North East Region of the country is an important region of genetic diversity which is being destroyed by jhumming, destruction of forests, soil erosion and other related causes such as developmental activities. It is therefore important to conserve the valuable genes by preserving valuable genes in original habitat or collecting the material before the man's greed destroyes the area. The genetic conservation is a complex process involving collection, storage and where possible preserving the original habitat.

The protein advisory group of FAO in its report (1973) recommended collection, maintainenance and evaluation of complete range of genetic diversity as it is of paramount importance for developing any meaningful programme of improvement and utilization of food legumes. It is further recommended that immediate steps be taken for systematic collection of legumes from centers of genetic diversity as well as from areas where from collections have yet to be made such as North Eastern Region of India.

The detailed investigations on Non-Conventional Legumes have been carried neither in India nor out side. The developed countries have concentrated their efforts on beans (Phaseolus species) and pea and have produced a large number of useful types both for canning and for use as fresh vegetables. The developed countries have also given a lead in finding new sources in proteins. The National Science Academy of United States constituted a panel in (1979) with the following objectives:

1. To identify leguminous plants currently little known and under utilized that could improve well being of people in developing countries

2. To select the species with the greatest inherent value
3. To indicate what needs to be done to capitalize on these plants and to develop them to their Potential

This panel produced a report entitled "Tropical Legume Resources for Future". This report lists about 200 species of little known legumes which can contribute to the economic well being of developing countries. Many of the species adapted to warm climate as we have in India particularly in North Eastern Region.

The above facts lead the author to write this book and so it is here.

LEGUMES OF NORTH EAST INDIA

Abrus prectorius Fabaceae

Common name : Kaunch, Chumhati

Distribution : All over N.E. Region in scrub jungles

Description: Woody climber probably profusely branching, glabrous or thinly silky. Leaflets, paripinnate in 10-20 opposite pairs, linear or linear oblong, minutely spiculate, glabrous above, silky beneath. Flowers reddish or dull white clustered in tubercles in dense pedunclate racemes.

Uses: Roots and seed are used medicinally. The seeds are used by Indian jewellers as lowest unit of weight called "Rati" weighing about 1.75 gms.

Abrus pulchellus Wall ex Thw. syn. **A fruticlosus** Mimosaceae

Common name : Kaich

Distribution : All over N.E. Region

Description : Twining shrub. Leaves paripinnate. Leaflets 16-30 oblong or obvate oblong apiculate, scattered adpressed, pubescent beneath. Flowers pinkish in fascicles of lax racemes. Corolla much exerted slightly adnate to staminal tube. Stamens 9, 10th absent.

Uses: Roots and leaves contain glycyrrhizin the principle constituent of liquorice and its decoction is given for coughs and colds. Roots diuretic tonic and emetic. Seeds administered in affections of nervous system.

Acacia catechu (L.) Willd. Mimosaceae

Common name : Khayer, Khair

Distribution : Assam and adjoining areas

Description : A middle sized decidous tree with rough brown or black bark, Pinnae 10-30 pairs leaflets usually 30-50 pairs. Flowers creamy white in axillary pubescent spikes.

Uses: The plant is a source of catechu (katha) obtained from heart wood of 20-30 year old trees. A gum is also obtained called khair gum and is used as a substitute for gum Arabic.

Acacia concina (Willd.) DC. Mimosaceae

Common name : Banritha, Latbabla, Ritha

Distribution : Found throughout N. E. Region

Description : A large scrambling shrub with conical and hooked prickles. Pinnae 3-6 pairs, leaflets 6-20 pairs, linear oblong, unequal. Flowers white with red median lines arranged in large panicles.

Uses: The fruit is used as detergent for washing hairs, woollens and silk fabrics. Tender acidic leaves used in chutney. Seeds are roasted and eaten. Decoction of pods used in biliousness. Bark used for tanning fishing lines.

Acacia farensiana (L.) Willd Mimosaceae

Common name : Bilati babul

Distribution : Fairly common all over N. E. Region

Description : A shrub with slender zigzag branches with grey dots. Pinnae 4-8 pairs. Leaflets 10-20 pairs, linear, oblong coriaceous. Flowers in heads, bright yellow fragrant, pods horizontally seriated.

Uses: Planted in gardens as hedge plant. The fragrant flowers are a source of perfume. The pods and roots are rich in tannins.

Acacia pennata Willd. Mimosaceae

Common name : Sujai-lewa, Bon-siri, Boloa, Serthing, Soh-mysiang, Suji-phang

Distribution : Assam and Meghalaya

Description : A large prickly climber, Bark dark brown to grey. Branches with about 5 rows of small or recurved prickles. Young shoots, Inflorescence, leaf rachis and pinnae pubescent. Leaflets from 16 pairs in lower pinnae to 30-50 pairs in middle pinnae. Leaflets linear-oblong, unequally broad, sharp and acute at end. Flowers pale white in pedunclate heads.

Uses: Bark is used as fish poison and it also contains tannins. Fruit pulp pesticidal, Decoction of leaves used as febrifuge.

Acacia pruinensis Kurz. Mimosaceae

Common name : Rampet-riubi, Te-bam, Suji-doukha, shitli

Distribution : Entire North East

Description : A large woody climber armed with recurved pricles, branches pruinose. Leaf rachis with a large gland, pinnae 8-15 pairs, leaflets 40-60 pairs closely crowded. Heads yellow arranged in axillary panicles.

Uses: Leaves and bark are medicinal. The leaves and stem are used as fish poison.

Acacia monliformis Griseb. Mimosaceae

Common name : N.A

Distribution : All over N.E. States cultivated as an ornamental tree.

Description: Medium sized tree with drooping branches. Bark grey brown. Leaves wanting. Phyllodes sub-coriaceous, nerves parallel. Flowers minute, yellow, scented in dense rigid spikes near the end of the branches.

Uses: Used as an ornamental tree.

Acacia mearnsii Willd. Mimosaceae

Common name: Black wattle, Schwarze-akazie, Gamboom

Distribution : Entire N.E. Region

Description : Tree 6-20 m tall, crown conical or rounded. All parts except flowers pubescent or puberlous. Leaves bipinnate, pinnae 8-30 pairs and pinnules 16-70 pairs. Flowers in globose heads borne in panicles or racemes.

Uses: Bark is one of the richest tanning material. Wood is used for paper and board manufacture. Spent bark as well as wood used for straw boards. It is best among tannin yielding Acacias. Used as astringent.

Acacia melanoxylon R. Br Mimosaceae

Common name : Australian blackwood

Distribution : Manipur in particular

Description : A moderate sized thorny tree 6-10 m tall. Leaves with 4-6 pinnae. Flower heads in few cm long spikes. Pods are long 6-8 seeded.

Uses: Wood beautifully mottled suitable for furniture. Tree yields a gum which is used in medicine. Bark contains 7-10% tannin.

Acacia nilotica (L.) Willd. Mimosaceae

Common name : Babul, Kikar, Egyptian mimosa

Distribution : All over N.E. Region in the lower parts of N.E. Region.

Description : A small tree with dark brown bark. Pinnae 3-6 pairs, leaflets 10-20 pairs membranous, Leaves often with 1-2 petiolar glands. Flowers yellow fragrant, spines grey pubescent slightly curved.

Uses: Bark is used for tanning. Gum obtained from the tree is used in confectionary also used for printing and dyeing. Seeds roasted and eaten. Decoction of bark used as a gargle and that of pods in urino-genital diseases.

Adenthera pavonina L. Mimosaceae

Common name: Kuchandana, Dieng-thong, Karli-arong

Distribution : All over N. E. Region

Description : A handsome middle sized deciduous tree. Bark dark brown or greyish brown. Leaves bipinnate, pinnae 3-6, leaflets 6-16 oblong to ovate-oblong or elliptic oblong, apex obtuse or sub-truncate, dark green and glabrous above. Flowers minute greenish yellow, Stamens 10 free, ovary many ovuled, Style filiform.

Uses: The seeds are strung into beads. They are supposed to weigh approximately 4 gms and are used as weight. A decoction of seeds and wood is used in pulmonary affections. Kernels contain pale yellow fat.

Aeschynomene aspera L. Fabaceae

Common name : Sola

Distribution : All over N.E. Region in swamps

Description : A little branched stout herb. Bark greenish grey, warty. Leaves decidous, leaflets many 60-100. Raceme lax flowered and often branched, clothed with bristly hairs. Flower yellow.

Uses : The wood is white pith like,soft and very light. It is used for hats and decorative material and also used as cork in bottles.

Aeschynomene indica L. Fabaceae

Common name : Kohila

Distribution : Common in the region in grasslands and upto 1000 m in hills

Description: A slender much branched annual, branchlets terete with minute warts. Leaflets 40-60, oblong, mucronate. Racemes axillary, appearing terminal after the leaves are shed, often branched. Flowers yellow.

Uses: Used as fodder but is toxic to horses in fruiting stage. Seeds yield a fatty oil.

Albizzia chinensis (Oesbeck) Merril. Mimosaceae

Common name : Shirish

Distribution : Assam and other tea growing areas in the region.

Description : A large decidous tree with flat crown. Bark dark grey, rugose. Leaves bipinnate, pinnae 6-20 pairs, leaflets 20-45 pairs, sessile, narrowly falcate, acute membranous. Heads in axillary and terminal panicled racemes. Flowers sessile, yellowish white. Calyx and corolla tomentose outside.

Uses: Wood is used for furniture and chest boxes. Used in tea gardens to provide shade.

Albizia julibrissin Duraz Mimosaceae

Common name : Gulabi Resham

Distribution : Manipur, Meghalaya, Nagaland, Sikkim upto a height of 1500 m.

Description : A medium sized deciduous tree, bark rough, dark brown or almost black, fissured blaze pink with distinct white streaks. Young shoots, rachis, pinnae and inflorescence covered with soft yellow or brownish pubescence. Pinnae 5-8 sometimes more upto 12 pairs, leaflets 10-

20 pairs. Flowers fragrant, yellowish green. Stamens exerted, rose pink in colour.

Uses: Leaves are a source of tannins. Seeds contain protein and oil. Bark is used in bronchitis, leprosy, paralysis and in helmenth infections. The bark and seed are astringent, useful in piles and diorrhoea and acts as restrorative.

Albizia lebbeck (L.) Benth. Mimosaceae

Common name : Sirisha, Rumri, Japhang, Dieng-sohrin, Amaltas.

Distribution : All over N.E. Region upto 1000 m. Planted as roadside tree.

Description : A large decidous tree with spreading crown. Bark light grey to dark brown or almost black. Branches lenticillate. Pinnae 2-5 pairs, leaflets sub-sessile 3-10 pairs. Heads pedunclate in terminal and axillary panicled racemes. Flowers sessile, yellowish white

Uses: The wood is used for high class furniture and internal decoration. The plants are reported to have antiseptic and anti dysenteric and anti-tubercular properties bark yields tannin. Leaves are nutritious and can be used as fodder. Leaves, flower and stem on distillation yield a colourless oil.

Albizia lucidor (Steud) Nielson syn. **Albizia lucida** (Roxb.) Benth. Mimosaceae

Common name : Sil-koral, kamatek, meji, kach

Distribution : All over N.E. Region ascending upto 1000 m.

Description : A middle sized nearly evergreen tree. Bark nearly smooth with horizontal wrinkles. Leaflets 2-3, rarely one pair. Heads small peduncled, corymbose sub umbellately arranged on branches. Flowers pale yellow.

Uses: Used for production of lac.

Albizzia myriophylla Benth Mimosaceae

Common name: Thabi-rekong, Mei-theit, Madhuron-glata

Distribution : Throughout N.E. Region.

Description : A large woody climber or a large shrub straggling on trees with raised dots on the branches. Pinnae 12-20 pairs, leaflets 25-50 pairs closely crowded, linear, often slightly falcate, base unequal, hairy along the margins. Heads small pedunculate, arranged in terminal leafy panicles. Flowers sessile. Filaments white or greenish staminal tube as long as corolla tube.

Uses: The fibrous bark tastes like liquorice.

Albizia odoratissima Benth Mimosaceae

Common name : Hiharu, Goroi, Cham-kaori, Turut-arong, Narlong, Khelbi, Ga.

Distribution : Throughout the N.E. Region.

Description : A large decidous tree with spreading crown. Bark grey or brownish grey to nearly black, rough outside, inside red with white streaks. Leaf rachis with large gland at the base. Pinnae 3-5 pairs, puberlous or glabrescent, leaflets 7-20, oblique oblong-obtuse or sub-acute, glabrous above glaucous beneath. Heads few flowered solitary or in fascicles of 2-5 on leafy corymbose panicles. Flowers sessile, fragrant, corolla funnel shaped hairy. Stamens pale yellow.

Uses: Wood is fairly resistant to white ants. Used for construction purposes and for agricultural implements. Leaves and twigs used as fodder. Also used as fish poison.

Albizia procera Benth. Mimosaceae

Common name : Korai, Jigring, bon phang, Sid-chachira-korai.

Distribution : Entire North Eastern Region upto 1000 M.

Description : A tall handsome decidous tree. Bark nearly smooth, yellowish, grey or brown outside pealing in irregular thin flakes. Leaflets 6-10 pairs, oblique oblong-ovate or rhomboid oblong, obtuse, silvery when young. Heads usually in fasicles. Sometimes solitary arranged in large terminal panicle, flowers sessile, yellowish white. Filament yellowish, tube as long as corolla tube

Uses: Suitable for posts and agricultural, implements and for construction.

Albizia stipulata Boivin Mimosaceae

Common name : Saw- Koroi, Sau, Horish, Shrisha, Arbhu, Bolphu, Borphi, Bhosra-koroi.

Distribution : Whole of N.E. Region ascending upto 2000m.

Description : A large handsome fast growing flat crowned tree. Bark smooth dark brown or grey outside, dark red or crimson with white streaks inside. Leaf rachis long with large gland near the base. Pinnae 6-20 pairs, leaflets subsessile 20-40 pairs, falcate oblong tapering to a sharp point, glabrate above pubescent beneath. Heads peduncled in terminal and axillary panicled racemes. Flowers sessile, yellowish white.

Uses : Wood used for tea and packing boxes. Leaves and twigs lopped for fodder. Gum is exuded which is used in sizing the paper.

Alysicarpus rogosus DC. Fabaceae

Common name : Shevra

Distribution : Manipur

Description : A prostrate sub-erect glabrous herb. Leaves oblong, rarely orbicular or linear lanceolate. Flowers reddish purple, compact spicate racemes.

Uses: A good fodder plant both green and dried. Fruits used as vegetable.

Alysicarpus tetragonolobus Edgew. Fabaceae

Common name : N.A.

Distribution : Assam, Manipur, Tripura.

Description : Small difuse shrub, herb. Stem slender. Leaves, unifoliate oblong. Stem hairy on one side with a line of adpressed hairs. Flowers -in axillary lax racemes. Pod tetragonous glabrous.

Uses : A good fodder plant.

Alysicarpus vaginalis (L.) DC. Fabaceae

Common name : Daval, Sangue

Distribution : All over N.E. Region.

Description : Robust herb. Leaves unifoliate, elliptic to lanceolate oblong or linear oblong. cordate at base, hairy beneath. Flowers in elongated racemes. Pods tetragonous, reticulate, slightly pubescent.

Uses: Good fodder, plant suitable for hay and makes good pasturage. Excellent soil improver. Root is used as anti fertility and as expectorant. Used in fracture, to hasten joining of the fractured part.

Atylosia goensis Dalz. Fabaceae syn. **Atylosia barbata** Baker

Common name : Rantur, Bankulthi

Distribution : Meghalaya (Khasi hills) upto 2000 m.

Description : A woody climber with pubescent terate branches. Leaflets 2-4 ovate-rhomboid and abruptly sub-cordate at the base, somewhat hispid above and softly pubescent beneath. Racemes terminally or axillary long elongated. Corolla long, yellowish. Pod compressed 5 seeded

Uses : Used as febrifuge.

Atylosia scarabaeoides (L.) Benth Fabaceae

Common name : Ban kalai, Barghore, Karbani.

Distribution: Nagaland, Manipur, Tripura and scattered in other parts of North East Region.

Description : Herb profusely branching, pubescent, branches twining. Leaflets subdigitate, strongly 3 nerved, terminal elliptic or oval oblong. Flowers yellow, on axillary racemes. Stamens diadelphous 9+1.

Bauhinia acuminata Caesalpiniaceae

Common name : Matikatota, Gai-chkong, Megong.

Distribution : Assam and adjoining areas.

Description : An erect herb, Young parts pubescent. Leaves slightly longer than broad, acute or subobtuse, glabrous above, finely pubescent beneath. Flowers white in short racemes opposed to leaves. Stamens fertile. Ovary spitate. Pod glabrous and beaked.

Uses: Decoction of bark and leaves is used in biliousness, leprosy asthma and stone in bladder. Flowers eaten raw. Seeds yield a fatty oil. Stem and bark are antidote to poison.

Bauhinia angunia Baker Caeasalpiniaceae

Common name : Nagput, Deo-jhakola, Dukhoi-Akhlem, Kapiadong, Tisodanari, Zongalee.

Distribution : Meghalaya, Assam.

Description : A large climber with flat ribbon like stem twisted to form depressions on alternate sides. Leaves broadly ovate, almost entire on old plants, deeply lobed almost to the base into acuminate or caudate lobes. Flowers very small in racemes arranged in terminal panicles often extending to the axils of upper leaves, white, puberlous. Stamens 3, Ovary stipulate, glabrous.

Uses : Source of fibre.

Bauhinia khasiana Baker. Caesalpiniaceae

Common name : N.A.

Distribution : Throughout N.E. Region.

Description : A large climber often deeply fluted. Leaves ovate or sub-orbicular lobes thinly coriaceous, glabrous, strongly 7-9 nerved. Flowers in corymb, several together forming a terminal spreading panicle. Flowers brownish red, Stamens 3, fertile Filaments tapering glabrous red. Ovary tapering densely pubescent.

Uses: Source of fibre and green manure.

Bauhinia malabarica Roxb. Caesapiniaceae

Common name : Kotra, Tenga kotra, Mugung, Laphang

Distribution : Assam, Meghalaya.

Description : A moderate size tree with bushy crown. Bark dark brown or grey. Leaves broader than long, cordate or sub-cordate, corieaceous glaucous beneath, glabrous above. Flowers dull white in axillary terminal racemes. Stamens 10, alternately shorter and longer. Ovary pubescent. Style protruding.

Uses: Young leaves and shoots are eaten. Used as medicine in blood dysentery.

Bahunia purpurea (Benth ex Baker) Blatter Caesalpiniaceae

Common name : Kunal, Kodra, Devchanchan, Kurrail.

Distribution : Assam kamrup, Goalpara, Lakhimpur.

Description : A moderate sized tree bark green somewhat rough in old stems. Leaves rather long than broad, glabrous above, base cordate 9-nerved, overlapping on the inner margin. Flowers purple in few flowered terminal corymbose or paniculate brown tomentose racemes, stamens usually 3, fertile, slightly shorter than the petals. Pods 6-10.

Uses : Heartwood reddish brown, moderately hard, suitable for agricultural implements. The leaves are used as cattle fodder. Bark used for tanning. Roots carminative. Bark used in diarrohea. Flower buds eaten as pot herb and also pickled. Leaf used in jaundice. Bark is also used in dropsy convulsions, pain, rheumatism and swelling.

Bauhinia tenuiflora Wall ex C.B. Clarke Caesalpiniaceae

Common name : Naga –hiamba, Jong-reckal

Distribution : North cachar hills, Jatinga, Haflong, Naga hills, Manipur.

Description : A largely cirrhose climber. Bark dark or reddish brown. Shoots and inflorescence clothed with copper coloured hairs. Leaves long usually as broad as long or even broader, thin glabrous above glaucous below. Flowers white in terminal and lateral many flowered corymbs. Fertile Stamens 3, ovary and pod glabrous.

Bauhinia variegata Linn. Caselpiniaceae

Common name : Tulla, Kanchan, Boga katra, Danchi-kasam, Bobmigang, Kural-bi-phang.

Distribution : Throughout N.E. Region upto a height pf 1000m.

Description : A moderate sized tree with dark-grey or brown, somewhat rough bark. Shoots tomentose. Leaves as long as broad or sometimes broader usually deeply cordate, thinly corieaceous glabrous and dull green above, grey glaucous and puberlous along the nerve beneath. Flowers large, pure white, pink or purple in few flowered corymose grey pubescent racemes. Style short.

Uses: The flower buds are curried and eaten. The leaves are good fodder. Often cultivated for its handsome Flowers. Bark used in diarrhoea and dysentery and also in leprosy, malaria and piles. Root is used in flatulence and indigestion. Leaf is laxative.

Butea buteformis (Voigt) Grieson and Long Fabaceae

Common name : Dieng-la-plang

Distribution : Meghalaya, Assam.

Description: An erect shrub with long switchy branches, covered with grey silky tomentum leaflets often upto 15 of more or less same shape. Racemes long erect, angular with often short branches at the base. Flowers long red.

Uses: Seeds in particular kill intestinal worms.

Butea monosperma (Lamk.) Taub Fabaceae

Common name : Palas

Distribution : All over N.E. Region.

Description : A middle sized decidous tree with crooked trunk and dark grey nodose branches. Leaves trifoliate, leaflets coriaceous, pubescent beneath, terminal obvate, rounded at apex. Flowers showy red on the leafless branches, petals 5, diadelphous 9+1, pod long.

Uses: Leaves used for making plates, cups etc. Dried leaves used as beedi wrappers. Seeds pounded with lemon juice act as a powerful rubifacient. Wood pulp is used for newsprint, antifertility, antihelmentic. Flower is aphorodiasic and diuretic. Leaf used on boils. Bark is used in bone fracture, diarrhoea and dysentery, Seeds are used as contraceptive and as vermifuge. Stem and bark act as fish poison.

Butea parviflora Roxb. Fabaceae

Common name : Salmana, Jong-obi, Baralokhar

Distribution : All over N.E. Region.

Description : A gigantic climber, Leaves trifoliate, elliptic or sub-rhomboid, cuspidate racemes in large terminal panicles. Flowers cream coloured or reddish. Stamens diadelphous. Pod one seeded.

Uses: Decoction of bark is given in dropsy and bowl complaints, root is also used in gum troubles,mucous in urine. Root is used in rheumatic and venereal disorders. Seeds yield a fatty oil used in cooking. It also yields a fibre which is used in ropes.

Caesalpinia bonduc (L). Roxb. Syn. **C. cristata**
Caesalpiniaceae

Common name : Lat kanta,Ghagargola, Karanj, Khar konda

Distribution : Entire N.E. Region particularly, Assam, Arunachal and Tripura.

Description : Scandent or climbing shrub armed with straight prickles. Leaves bipinnate leaflets 6-10 pairs, oblong or elliptic, mucronate, puberulous. Racemes supra axillary and terminal, peduncled. Petals yellow sometimes spotted with red. Filaments hairy.

Uses: Cultivated as hedge plants. Leaves are used against tapeworm, also used in malaria, intermittent fever, pulmonary tuberculosis, mensurational complications, rheumatism, skin disorders and wounds. Seeds yield medicinal oil.

Caesalpinnia digyana Rotler Caesalpiniaceae

Common name : Umul-kuchi

Distribution : Probably all over N.E. Region.

Description : A large scandent or straggling shrub armed usually with hooked pricles. Pinnae 8-12 pairs. Leaflets 7-10 pairs in each pinnae, oblong glabrous on both surfaces, apex obtuse or rounded, base oblique. Racemes supra-axillary and terminal, sometimes branched, pubescent when young. Flowers bright yellow, Filaments duly exserted lower portion densely wooly.

Uses: Pods contain much tannin. Seeds used as cattle feed mixed with other seeds.

Caesalpinnia mimosoides Lamk. Caesalpiniaceae

Common name: Kila sandu, timmullu

Distribution : All over N.E. Region.

Description : Shrubs, prickly, downy. Leaflets 10-20 pairs membranous glabrous raceme rachis densely prickly. Flowers bright yellow.

Uses : Used in ulcers and sores.

Caesalpinnia pulcherrima (L.) Swartz Syn.
C. capitata (Roth) Alston Caesalpiniaceae

Common name : Krishna-cheira, Peacock flower

Distribution : All over N.E. Region.

Description : A handsome branching shrub. Leaves with 6-9 pairs of pinnae. Leaflets 8-12 pairs, broadly oblong or obovate-oblong, emarginated and mucronate glabrous. Flowers orange or yellow in elongate terminal and axillary racemes.

Uses : Cultivated as an ornamental plant. Whole plant is used in jaundice. Flowers used in body ache. Seed is used in gum trouble and ring worm.

Callindra haemotocephalla Hask Mimosaceae

Common name : Diang-si-soh-maria, Khi utilou: Manipur: Palpia, Naga, Thing sakai.

Distribution: Assam, Meghalaya, Nagaland, Manipur, Tripiura.

Description : Small tree, Leaves bipinnate, leaflets few large, stipules persistent. Flowers in globose heads. Corolla funnel shaped. Stamens indefinite ovary many ovuled. Stigma minute, capitate.

Callindra umberosa (Wall.) Benth Mimosaceae

Common name : Dieng sia soh-maria, Dieng-sia-soh-krot; palpani, Soumbu-baprong, thing-Seki

Distribution : Assam, Meghalaya.

Description : Usually small tree, sometimes middle sized, Stem with brown bark. Leaves bipinnate, pinnae one pair terminal and usually with a gland between them, leaflets in each pinnae. one pair terminal obliquely oblong-elliptic, lanceolate, bluntly acuminate, glabrous, rigidly sub corriaceous heads peduncled, globose/usually clustered on dead wood. Flowers yellowish – white, Pod valves finely reticulate. Seeds 4-9.

Uses: The wood burns with offensive smell.

Canavalia ensiformis (Linn.) DC. Fabaceae

Common name : Makhan Sem Kamtai- wahai, mepak-rekang,, Happa-tepi

Distribution : Assam.

Description : A fairly long perennial climber. Twining from right to left. Leaves pinnatele tri-foliate, leaflets rhomboid elliptic. Racemes axillary on terete peduncles. Flowers, large, white,usually with a dash of pink or purple, solitary. Stamens monodelphous, Ovary ovuled, Pod flat, Seeds 6-10,white.

Uses: Cultivated for forage and green manure. Very young pods used as Snap beans.

Canavalia gladiata (Jacq.) DC. Fabaceae

Common name : Bara sem. Makhan Sem

Distribution : All over N.E. Region.

Description : A large perennial climber, Leaves pinnately trifoliate, ovate-rhomboid elliptic, terminal largest. Racemes axillary on long peduncles. Flowers large, white with purple tinge. Stamens 10, Seeds large red.

Uses : Seeds edible, Root is used in intestinal worms.

Cassia alata Linn. Caesalpinaceae

Common name : Dadulph, Daudpata

Distribution : All over N.E. Region.

Description : A soft wooded shrub with thick downy branches. Leaflets 10-12 pairs, oblong, obtuse but apiculate with unequal base, glabrous on both surfaces or slightly downy beneath. Racemes speciform, stout. Flowers bright yellow. Stamens 7, fertile 2 longer than others.

Uses: Leaves are used in ringworm. Seeds used as vermifuge.

Cassia fistula L. Caesalpinaceae

Common name : Sonaru, Sonari, Honaru, Sonaru arang, Sonaru phang, Sonali

Distribution : Throughout N.E Region upto 1000 m.

Description : A medium sized tree. Bark pale or ash grey. Leaves long, rachis puberlous leaflets 4-8 pairs, opposite ovate elliptic-oblong. Flowers yellow in axillary lax drooping racemes, pubscent when young, gradually larger towards base. Stamens 10, 3 lowest longest with curved filaments indehiscent anthers.

Uses: Bark known as sumari is used as admixtrure with avaram bark for tanning.

Cassia floribunda Cav. syn. **C. corymbosa, C. laevigata, C. quaderangualis** Caesalpinaceae

Common name : N.A.

Distribution : Meghalaya.

Description : A handsome shrub with glabrous branches, Leaflets 3-5 pairs lanceolate. Flowers yellow in peduncled corymbose racemes. Stamens 7, fertile 2 longer than others, 1 intermediate and 4 short.

Uses: Leaves and bark used in tanning.

Cassia javanica L Caesalpinaceae

Common name : Morai, Bandar lathi, Choui, Khordonu-Khadong

Distribution : All over N.E. Region.

Description : A small middle sized tree with pubescent shoots, Bark grey. Leaves long leaflets 10-12 pairs. Flowers pink in corymbose racemes, Stamens 10 short, 3 long, 3 incurved bearing sterile anthers.

Uses: Used as fodder.

Cassia nodosa Ham. Casealpiniaceae

Common name : Moroi, Bandar lathi, Khorina-khodong

Distribution: Arunachal, Assam, Meghalaya & Manipur

Description : A small to middle sized tree with pubescent shoots. Bark grey or dark brown yellowish near the cambium. Leaves decidous early leaflets 10-12 pairs lowest usually smaller, oblong lanceolate or elliptic-oblong sometimes bluntly acuminate. Flowers pink in corymbose raceme stamens 10. Filaments yellow, 3 longer with globose swellings 4 shorter apparently sterile anthers.

Uses: A handsome tree when flowering.

Cassia occidantalis L. Caesalpinaceae

Common name : Hant-menga, Naguni, Jhar-jhani-gach, Mithep-phang, Hadi-dika-darong, Taw-Eit-niang-thynthai, Ponthagarai, Rautakla, Chek-unda

Distribution : Common through N. E. Region upto 1000 m.

Description : A foetid soft wooded undershrub, Leaflets 4-5 pairs ovate to oblong lanceolate sharply acuminate. Flowers yellow in few flowered short axillary or terminal corymbose raceme. Stamens 10, 3 reduced to staminodes. Seeds 20-30.

Uses : The seeds are said to be used as substitute for coffee in Africa and Mauritius the plant is also used for skin diseases. Leaf is diuretic used in dropsy, dysentery eczema, fever, gastric complication. Root is used for ringworm, skin disorders and rheumatism

Cassia renigera Wall ex Benth. Caesalpinaceae

Common name : N.A.

Distribution : All over N.E. Region.

Description : Small decidous tree with spreading and drooping branches. Leaflets 10-20 pairs almost sessile oblong obtuse rounded at the base. Flowers pinkish scented in short racemes. Stamens 10 with cylindrical thickning in middle of filaments.

Uses: Planted as an ornamental plant.

Cassia siamea Lamk. Caesalpinniaceae

Common name : Minjori

Distribution : Tripura.

Description : A middle sized tree, wood brown to black. Leaves without glands Leaflets 6-12 pairs, emarginated, mucronate. Flowers bright yellow in large terminal panicles of corymbose racemes.

Uses: Very fast growing, good as soil binder, wood hard used for miscellaneous purposes.

Cassia sophera Linn. Caesalpinaceae

Common name : Kalaksunda

Distribution : All over N.E. Region.

Description : An undershrub about 1 M tall. Younger parts pubescent. Leaf rachis with a conical gland at base. Leaflets 6-12 pairs lanceolate or narrow lanceolate, acuminate or acute base somewhat unequal glaucous beneath. Flowers yellow in short axillary and terminal corymbose racemes panicled at the end of branches.

Uses: Leaves, bark and seeds cathartic. Leaf juice specific for ring worm.

Cassia timorena DC. Prodr. Caesalpiniaceae

Distribution: Tripura, Mizoram, Manipur & Meghalaya

Description : Small trees. Leaves without gland, leaflets 10-20 pairs oblong, obtuse mucronate, pubescent. Flowers in axillary corymbose racemes in terminal panicles pink to purple.

Uses: Similar to *Cassia tora*.

Cassia tora L. Caesalpinaceae

Common name : Bin-medilina, Dan-diga, Taw-eit-nyang-synrang, Taw-blei, Hadi-dika-arong

Distribution : Assam, Meghalaya upto 1000m.

Description : An annual attaining about 2 m with pubescent young parts. Leaflets 3 pairs obvate or obvate oblong, obtuse or sub-acute. Flowers yellow borne in pairs on very short axillary peduncles. Stamens 7, perfect 3 reduced to staminodes

Uses: The leaves and seeds contain chrysophonic acid and are good remedies for skin diseases like ring-worm.

Clitoria ternate L. Fabaceae

Common name : Aparajita

Distribution : All over N.E. Region as an ornamental plant.

Description : A trailing ornamental plant with white or blue flowers 5-7 leaflets. Flowers solitary in short peduncles. Pods long with 5-10 seeds.

Uses: Cultivated as an ornamental plant. Root is used in goiter, leprosy, snake bite, as purgative and stomach ache and abdominal swelling. Leaf is used in swelling. Seed is laxative.

Crotolaria albida Heyne ex Benth Fabaceae

Common name : Banmethi

Distribution : All over North East Region.

Description : Herb or undershrub with numerous slender branches. Leaves simple linear oblong or oblanceolate obtuse or mucronate. Flowers few in long racemes.

Uses: Root is used as purgative.

Crotalaria acicularis Buch Ham ex Benth. Fabaceae

Common name : N.A.

Distribution : All over N.E. Region.

Description : Small herbs diffuse with woody root stock. Leaves simple, elliptic or orbicular-oblong sub-cordate unequal at base. Flowers yellow in densely silky hairy racemes with short peduncles.

Uses : Leaves and roots are medicinal.

Crotalaria assamica Benth Fabaceae

Common name : N.A.

Distribution : Assam and Meghalaya.

Description : A shrub with stout branches which are finely striated and silky. Leaves oblanceolate-oblong, apex obtuse, subacute. mucronate, thinly corieaceous Racemes terminal about a foot long, 20-30 flowered. Flowers golden yellow.

Uses: Leaves and roots are reported to have medicinal properties.

Crotolaria bilata Schrank Fabaceae

Common name : Jhunjhuni, Ghurghuri, veersum

Distribution : Entire N.E. Region common in grasslands and along the roads.

Description: Undershrub, 50-60 cm long, silky, pubescent, winged, Leaves simple subsessile, ovate or ovate oblong. Oblique at base. Racemes few flowered. Flowers yellow.

Uses : Whole plant facilitates child birth, snake bite, tonic to pregnant women, also used as fodder. Root is used in urine complications.

Crotolaria calyciana Schank. Fabaceae

Common name : N.A.

Distribution : Tripura in particular but also found in other parts of region.

Description : Herb about 1 m tall. Leaves simple, subsessile, acute. Flowers axillary or in terminal racemes, yellow.

Uses: Used in carbuncle, cholera, convulsions, dysentery, haematuria, Venereal sores.

Crotolaria ferrugieniea Grah ex Benth Fabaceae

Common name : N.A.

Distribution : All over N.E. Region particularly upper Assam. Manipur and Meghalaya upto 1000 to 2000 m.

Description : A much branched robust herb. Branches conspicuosly with persistent decurrent stipules. Leaves simple sub-sessile obvate-oblong, obtuse or rounded at the apex. Racemes long peduncled, laxly 2-8 flowered. Flowers yellow. Pods linear oblong, 20-30 seeded.

Uses: Foliage is fodder for sheep.

Crotolaria humifusa Grah ex Benth Fabaceae

Common name : N.A.

Distribution : Common in Megalaya in grasslands upto 2000 m.

Description : A much branched slender herb clothed with dark brown silky hairs, leaves simple nearly sessile. Oblong, glabrous beneath. Racemes 3-6 flowered. Yellow.

Uses: Used in sores and wounds.

Crotolaria juncea Linn Fabaceae

Common name : San, Ausa, Sanmai

Distribution : Throught N. E. Region.

Description : A shrub 1-2 m tall. Leaves linear to obovate oblong, obtuse or sub-acute racemes simple mostly terminal, lax, 10-12 flowered. Flower bright yellow.

Uses: Bark yields a strong fibre more enduring than jute. Used for cordage, canvass and fishing nets. Also used as green manure, Abortif, anti-fertility, blood Purifier cough and skin diseases.

Crotolaria mysorensis Roth. Fabaceae

Common name : Lac soh, Mynkhar

Distribution : Throught N. E. Region specially in hill region upto 2000-3000 m.

Description : A densely diffuse herb. Leaves simple, oblong-elliptic-ovate racemes terminal or sub terminal long peduncled. Flower yellow.

Uses: Reported to have medicinal properties.

Crotolaria occulata Benth Fabaceae

Common name : N.A.

Distribution : Common in grass lands in hills upto 2000-3000 m.

Description : A small undershrub 1 m high with ascending branches. Leaves simple cuneately or oblanceolate. Racemes terminal, peduncled. Flowers 6-20 flowers blue with darker blue or purple veins.

Uses : Cultivated in gardens for its scented flowers.

Crotolaria pallida Ait: Fabaceae

Common name : Runcha, Thung-kur

Distribution : All over N. E. Region in wasteland and along roadside.

Description: Undershrub with slender sulcate branches, Leaves trifoliate, leaflets ovate-oblong or elliptic, mucoranate. Racemes terminal & lateral, Flowers yellow, streaked with purple.

Uses: Cover crop and for green manuring, Root is used in swelling joints also vermifuge.

Crotolaria prostrata Rottler ex Willd Fabaceae

Common name : Chota jhunjhunia

Distribution : Tripura and other parts of N.E. Region.

Description : A diffuse herb with perennial root stock, branches slender with silky hairs. Leaves sub-sessile, ovate-oblong or obvate,obtuse oblique at base silky above glaucous beneath. Racemes on slender peduncles, 2-4 flowered. Corolla small yellow.

Uses : Used as medicine after child birth.

Crotolaria sesiflora L. Fabaceae

Common name : N.A

Distribution : Along water courses or margin of swamps particularly in Meghalaya.

Description : A robust perennial upto 3 m with ascending strigose branches. Leaves simple, linear lanceolate, narrowed at both ends. Flowers in raceme or solitary in the axils of leaves. Flowers bluish white.

Uses: Leaves and roots are reported to have medicinal properties.

Crotolaria spectabilis Roth. Fabaceae

Syn. C. cuneifolia, C. leschehaulti, C. macrophylla, C. retz, C. sericea

Common name: Pipali-jhanjha, Jhu-jhumaria, ghantkaran

Distribution : Whole N.E. Region.

Description : Robust undershrub, branches short striated. Leaves broad, oblanceolate to obvate glabrous above pubescent beneath. Racemes lax peduncled. Flower yellow tinged with purple.

Uses: Used as green manure. Stem is anthelmintic and used in scabies. Whole plant is used as an antidote in poisoning.

Crotolaria tetragona Andr. Fabaceae

Common name : N.A.

Distribution : Grasslands in plains and lower hills in the drier parts of the region.

Description : A shrub upto 2 m in height tetragonous and sulcate branches. Leaves linear or lanceolate very gradually narrowed to cuspidate tip, racemes mosly terminal, lax 4-10 flowered. Flowers handsome yellow. Corolla glabrous slightly exerted.

Crotolaria verecusa Linn. Fabaceae

Common name : Bansan

Distribution : Entire N.E. Region

Description: Tall undershrub, Leaves ovate rhomboid, pubescent beneath. Flowers bluish purple in long axillary racemes.

Uses : Leaves used for scabies.

Dalbergia lanceolaria L Fabaceae

Common name : Kurakada, Neri-mungo, Patradi

Distribution : Upper regions of North East.

Description : A large decidous tree sometimes buttressed at the base, bark smooth, leaflets 7-17, elliptic oblong rounded at the base. Panicles axillary and terminal. Flowers dull white or pinkish. Stamens in two bundles of 5 each.

Uses : Wood used for agricultural implements, hand tools, carving and rafters. Decoction of bark is used in dyspepsia. Bark contains tannins, stem and bark used in fever. Stem is used in tongue infection and as tooth brush. Seed oil is used in rheumatism.

Dalbergia rimosa Roxb. Fabaceae

Common name : Gajai-lata, Mermeseri-lata, Kata-ghoh, sibou-daukho, Sibou-phang, Mei-nup-synrong, Jerma-bu-tem, Dieng-thiniat-nerai.

Distribution : All over N. E. Region.

Description : A sub erect, straggling or trailing shrub. Bark white outside, yellow inside. Leaflets elliptic, obvate or oblong acute or rounded at base. Panicles axillary or terminal with corymbose branches, Flowers very small say the smallest in Fabaceae, dull white, sweet scented.

Uses: Stem is used as handles. Seeds are eaten by tribals of Assam.

Dalbergia sissoo Roxb. Ex DC Fabaceae

Common name : Mihu, Shisham, Khujrap

Distribution : All over N.E. Region in sub-Himalyan tract.

Description : Ordinarily middle sized gregarious tree but in favourable localities becomes a big tree. Leaflets 3-5, distinctly alternate, ovate to obvate orbicular or rhoimboid. Flowers in cymose racemes which are corymbosely arranged in axillary peduncled panicles, yellowish white. Stamens 9 which are united in a sheath which is slit along the top, ovary pubescent, style short.

Uses: Valuable for construction of furniture and doors etc. Leaves used for fodder leaves, are anthelmintic, also used in dysentery, eruptions and gonorrhoea and skin diseases. Stem decoction is used in headache and leprosy.

Dalbergia stipulacea Roxb. Fabaceae

Common name : Dat-bijla or bijuli, Garo-deosal, Phawang, Khot bandu, Sabima vikang, Dieng-me-so-khwang, Chamengpe-sude-arong

Distribution : Throughout in the plains of N.E. Region but very common in Garo hills.

Distribution : A large climbing shrub often small tree or bush. Leaflets 17-25, cuneate and slightly oblique at the base, emarginate, membraneous. Panicles axillary ascending more or less secured racemose branches, Flowers rather distinct purple. Stamens delphous in two bundles of 5 each.

Uses: Small posts and tool handles. Bark and stem used as fish poison.

Dalbergia thomsoni Benth Fabaceae

Common name : Bon- scibai-doukha, Subin-rikang

Distribution : Present in Khasi District of Meghalaya and Cahar of Assam but not very common.

Description : A large woody climber, glabrous throughout. Leaflets often sub-opposite 9-15 oblong-obovate or elliptic, emarginated more or less cuneate at base flowers very small falling off very early. Pod thin but firm, greenish glabrous, thickened and veined over the solitary seed.

Uses: Leaves are used as fodder.

Dalbergia volubilis Roxb Fabaceae

Common name : Budu-galwang

Distribution : Entire North Eastern Region.

Description : A large woody climber with deeply fluted ash coloured stem. Branchlets often functioning as tendrils. Leaflets 9-13, obovate. Cuneate apex obtuse, truncate or emarginated, glabrous and green above pubescent and somewhat pale beneath. Panicles terminal and axillary. Copious, densely rusty pubescent. Corolla bluish white or light mauve. Stamens diadelphous in two bundles of 5 each.

Uses : Leaves used as fodder and their juice for gargle.

Dalhhousiea bracteata Fabaceae

Common name: Paliari-lata, Teli-lata Pantha-gach Mei-dieng, Luyang-tha-kali-gauj

Distribution : All over N.E. Region.

Description : An evergreen woody climber with ashy dark bark. Young shoots have thin hairs. Leaves bifarious, unifoliate. Leaflets ovate-elliptic or oblong elliptic, quite entire, abruptly bluntly acuminate, rounded or sub-cordate. Flowers bracteate,white. Stamens 10, filaments free.

Uses : The leaves are excellent as wrappers for beeris.

Delnox regia (Boj.) Raf. Caesalpiniaceae

Common name : Gulmohar, Krishnachura

Distribution : All over N.E. Region. Grown as an avenue tree.

Description: A medium sized tree with spreading crown. Leaves bipinnate, leaflets many. Flowers large crimson streaked with red and yellow borne in terminal corymbs.

Uses: Grown as avenue tree. Seeds contain gum which may find use in textile and food processing industries.

Derris robusta (Roxb. ex DC.) Benth Fabaceae

Common name : Mouhita

Distribution : Tripura and other N.E. States.

Description : Middle sized trees, deciduous. Leaflets 7-23 opposite, elliptic-oblong to oblong-lanceolate mucronate, thinly corriaceous, glabrescent above, minutely silky beneath. Flowers white, fascicled in axillary raceme. Seeds brown compressed.

Uses: Wood used for tea chests, post and ploughs. Leaves lopped for fodder.

Derris trifoliata Lour. syn. **Pongamia pinnata** Fabaceae

Common name : Karanja, Kanji

Distribution : Assam, common in costal forests.

Description: A large evergreen, glabrous climbing shrub.

Uses : Used as stimulant, anti-spasmodic, and counter irritant. Leaves are used as fodder. Stem is used in preparing rough cordage. Seed decoction is used in bronchitis, in burns and chest pain. Seed oil is used in whooping cough and cold. Flowers are used in diabetes and also in eczema.

Desmodium alatum DC Fabaceae

Common name : N.A.

Distribution : Throughout N.E. Region.

Description: Shrub 1-2 m high. Leaves narrow. Racemes densely scattered. Flowers small, light pink with dark purple veins at the base of petals.

Uses: Used as soil binder and grazing.

Desmodium caudatum (Thunb.) DC. Fabaceae

Common name : N.A.

Distribution : Throught N.E. Region.

Description : Shrubs slender with terete branches. Leaves trifoliate, leaflets oblong-lanceolate, acute, glabrous above pubescent beneath. Racemes long, lax. Flowers whitish.

Uses: Used as soil binder.

Desmodium concinnum DC. Syn. **D. retusum** Fabaceae

Common name : N.A.

Distribution : Common in Meghalay from 1000-2000 m.

Description : A handsome shrub with slender terete drooping branches. Young parts grey pubescent, Leaves trifoliate covered with spreading hairs, leaflets oblong, oblong-obvate or elliptic-oblong, rounded at both ends mucronate, green and glabrate above and clothed with grey hairs beneath. Racemes terminal and at the end of lateral branchlets or axillary and lax. Flowers pink or lilac.

Uses : Used as fodder and soil binder.

Desmodium ferrugineum Wall ex Thaw Fabaceae

Common name : Bae-phool, Krishnanupani

Distribution : Tripura and other parts of region.

Description : An erect undershrub 1 m high.

Uses: Plant has been tried as green manure and cover crop in coconut plantation. Also used for planting on terraces and on undulated land, prevents soil erosion. Decoction of the plant is considered as tonic and also used in cough.

Desmodium gangeticum (L.) DC. Fabaceae

Common name : Nai-khong, Salpan or Saluani, Salaparni

Distribution : Common in sub-montane region upto 1000-1500 mt.

Description : A undershrub 1 m tall, branches subterete, grey downy towards the extremeties. Leaves unifoliate. Leaflets ovate to lanceolate or elliptic or elliptic lanceolate, acute, rounded at the base. Racemes mostly terminal often with short lateral branches rather lax, pubescent. Flower purplish white.

Uses: Root is used as febrifuge, as an antidote to snake-bite, asthma, dysentery fever, vomiting. Whole plant is used in bronchitis, and as tonic. Stem fibre is used as a raw material for paper making.

Desmodium griffithianum Benth Fabaceae

Common name : N.A

Distribution : Common in Meghalaya upto 1000-1500 m and also throughout N.E. Region in abandoned jhum lands.

Description : A trailing herb ½ m long shaggy with fine spreading hair. Leaves trifoliate, leaflets terminal largest, obvate or oblong, glabrous above hairy beneath. Racemes terminal dense towards the top. Flowers bluish purple. Pod integuments 4-5 obscurely downy

Uses: Plant as a whole is medicinal.

Desmodium gyroides (L.) DC. Fabaceae

Common name : N.A.

Distribution : Throughout Assam.

Description : A shrub 2-4 m tall. Stem freely branched, grey hairy when young. Leaves normally trifoliate but often unifoliate by suppression, leaflets ovate-elliptic or obvate, obtuse. Racemes axillary and terminal, usually branched, very hairy elongating in fruit. Flower purplish. Stamens diadelphous. Seeds reniform with a winged strobile, black.

Uses: Leaves made into poultice and applied in Lumbago. Plant promotes conception.

Desmodium heterocarpon (L.) DC. Fabaceae

Common name : Sarivan

Distribution : Tripura

Description : Undershrub, branches grey tomentose Leaves trifoliate, leaflets sub-equal, ovate, elliptic, oblong, glabrous above, pubescent beneath. Flowers in terminal and axillary racemes

Uses: The whole plant is used in cough, fainting and tonic. Root used in fever.

Desmodium heterophyllum (Thunb.) DC. Fabaceae

Common name: Khaldi, Kunnapapalai, Pookarisa, Tinkoriamioni.

Distribution : Meghalay both in hills and valleys and usually in meadows.

Description: Large shrub. Leaves trifoliate, leaflets obvate-oblong hairy beneath. Flowers purple, axillary solitary or two together or 2-6 flowered in lax raceme. Pod much indented along the lower suture, joints 4-5 as broad as long.

Uses : Roots carminative, tonic emegouge, diuretic.

Desmodium laxiflorum DC. Fabaceae

Common name : Kihong-sir, Kadakatru

Distribution : Assam, Meghalaya.

Description : An erect undershrub about 1 m tall branches angled. Leaves trifoliate, leaflets terminal largest lateral, somewhat oblique, ovate, ovate-elliptic, or lanceolate, cuspidate-acute or sharply acuminate base rounded, membranous green and scabrid. Racemes axillary and terminal sometimes panicled, laxi-flowered. Flowers fascicled, purplish pink.

Uses : Smallpox.

Desmodium motorium (Houtt.) Merr. Fabaceae

Common name : Gorach, Bhuncara

Distribution : Fairly common in grasslands

Description : A small undershrub, branches striated. Leaves trifoliate, oblong, lanceolate, obtuse at both ends glabrous above silky beneath. Flowers pink in axillary and terminal racemes often panicled.

Uses : Rheumatism and wounds.

Desmodium pulchelum Fabaceae

Common name : Jula, salpani, chapor, dehkna nandek, jeetdari

Distribution : Common in decidous forests.

Description : A shrub with slender terete or ribbed, grey downy branches. Leaves trifoliate, leaflet terminal largest, ovate-lanceolate Or elliptic-oblong, obtuse at apex rounded at base, coriaceous glabrous above finely downy beneath. Flowers in clusters of 2-6 in erect terminal axillary racemes. Flowers white or tinged with pink.

Uses : Bark and flowers have medicinal value. Hypotensive. Used in Dental cavities and headaches.

Desmodium triquetrum DC. Fabaceae

Common name : Ulucha, Ettang

Distribution: All over N.E. Region in planes upto 1500 m.

Description: An erect or-sub-erect undershrub generally 1 to 1.5 m tall but sometimes attaining a height upto 2-4 m. Branches shortly triquetrous and strigose along the ridges. Leaves unifoliate, leaflets oblong-lanceolate, rounded or sub-cordate at the base acute or shortly acuminate, entire, sub-coriaceous with scattered bristles like white hairs. Racemes axillary or terminal and sometimes panicled covered with minute pubescent and with stiff adpressed hairs. Flowers purple, Stamens monodelphous

Uses: Used as tea leaves by some tribes. Extract of leaves used in piles.

Desmodium velutinum (Willd.) DC. Fabaceae

Common name : Salpani, Latkakani

Distribution : Common in open grassland and thicket.

Description : An undershrub, branches terete. Leaves unifoliate, leaflets ovate or deltoid, subobtuse, mucronate truncate or cordate at the base, hispid above, brown tomentose beneath. Racemes axillary or terminal. Flowers fascicled purple.

Uses: The root has medicinal value.

Dolichos uniflorus Fabaceae

Common name : Pataluri

Distribution : Cultivated in domestic compounds

Description : Annual climbing slender herb densely hairy. Leaves pinnately trifoliate, leaflets ovate to elliptic to rhomboid cuneate at the base, sub-acuminate or acute at apex. Flowers 1-3 axillary, yellow.

Uses: Burns, dysentery, kidney stones, smallpox, measles, Pneumonia.

Dunbaria rotundifolia Weight & Arn. Fabaceae

Common name : Kunjaban

Distribution : Found at the outskirts of forests.

Description : Herbs filiform, twining with downy stem Leaves pinnately trifoliate, leaflets orbicular rhomboid, membranous above downy gland dotted beneath. Flowers soitary or few on a Short peduncle reddish in colour. Stamens diadelphous 9+1.

Uses: Used as pulse and fodder.

Entada phaseoloides (L.) Merr. Fabaceae

Distribution : All over N.R. Region

Description: A large climber with blackish bark. Leaves bipinnate rachis ending in a long bifid tendril, leaflets 2-4 pairs obvate, oblong or elliptic, emarginated, oblique at base coriaceous. Flowers scented axillary or panicled from uppernodes of old leaflet branches.

Uses: Eaten after roasting. Seeds also used as substitute for soap.

Erythrina arborescens Roxb Fabaceae

Common name : Dieng-son, Dieng rsong-ja-ra-song, Dieng-r-song-hadem, Dieng-song.

Distribution : Meghalaya in particular upto 2000-3000 m and also in other parts of N. E. Region.

Description : A small tree with low spreading crown. Bark rough with a shining layer of cork outside bright green underneath the cork inside pale. Leaflets broadly triangular-ovoid or rhomboid with a cuneate base sharply acute. Racemes axillary long peduncled pubescent when young. Flower orange red.

Uses: Seeds yield a fatty oil. Bark and leaf are used in skin diseases. Wood is used in drums. Plant is used as hedge in garden

Erythrina fusca Lour. Syn. **E. glauca** Willd., **E. ovalifolia** Roxb. Fabaceae

Common name : Kanta, Madar, Harikakra

Distribution : Entire N.E. Region

Description : A small tree, branches armed with numerous black pricles. Leaflets ovate oblong sub-coriaceous glabrous above glaucous beneath racemes terminal, Flowers scarlet red mostly in groups of 3 on rachis

Uses: Planted as hedge plant and to support betel vines.

Erytherina subumbrans (Hassk) Merr. Fabaceae

Common name : Madar

Distribution : Tripura and other adjoining states.

Description : A large decidous tree. Leaflets glabrous membranous, ovate-roundish, acuminate. Racemes pilose, Flowers red.

Uses: Used as green manure

Erythrina stricta Roxb Fabaceae

Common name : Madar,, Tagat-asing, Pharche-arong, Mandal, Antu, Bag-long-thing

Distribution : Fairly common in N.E. Region upto a height of 1000 m

Description : A fairly large tree with a striking appearance when in flower. Bark deeply furrowed greenish yellow outside. Leaflets sometimes broader than long rhomboid-orbicular or ovate or shortly acuminate, glabrous and shining above, glaucous and pubescent beneath. Racemes horizontal crowded at the end. Flowers coral red.

Uses: Wood is used as planks. Bark is used in fever, mensturational complications stomach diseases and snake bite. Leaf is used as tonic and vermifuge.

Erythrina variegata L. Syn **E. indica** Lamk, **E. mysorensis** Gamble, **E. picta** Lour Fabaceae

Common name : Kanda madar, Paldha,

Distribution : All over N.E. Region

Description : A decidous tree, prickly with yellowish bark, leaflets glabrescent, terminal roundish, cuspidate, truncate or rhomboidal at base. Racemes dense. Flowers scarlet red.

Uses: Cultivated as hedge plant in gardens. The tender pods are edible. Bark fibre used as cordage. The flowers yield red dye. Bark and leaf is used in cold and cough, eye complications, fever, skin diseases, ulcer, snake bite, Whole plant is used in convulsions.

Flemingia bracteata (Roxb) Wight syn. **F. strobilifera, Moghania bracteata** Fabacea

Common name : Makhioti

Distribution : All over N.E. Region upto 1000-1500 m

Description : A much branched scandent shrub 3-4 m high. Branches terete, pubescent in young plants. Leaves unifoliate, ovate lanceolate. Flowers white or pinkish covered by large folded bracts.

Uses: The roots are used in epilepsy and hysteria. Also given to relieve pain and induce sleep also used as vermifuge for childern.

Flemingia fruticulasa Wall ex Benth. Fabaceae

Common name: Dukhandada, Kanphuta, Nappie,, Surring-sarong

Distribution : Found in grasslands of N.E. Region upto a height of 1500 m

Description : Undershrub, Branches trailing, striate, grey, tomentose. Leaves unifoliate ovate or elliptic cordate at the base acute or oblong at apex, Racemes terminal. Flowers white.

Flemingia involucrata Wall ex Benth. Fabaceae

Distribution : Occurs in N.E. Region but rarely

Description: An erect shrub about 1 m tall. Leaves tri-foliate, leaflets, elliptic oblong lateral oblique chartaceous. Heads many flowered both terminal and axillary. Flowers purple. Pod more on less papery, Seed dark brown or almost black

Flemingia macrophylla (Wlld.) Kuntze Fabaceae

Common name : Samnaskhat, Bhajangabasa, Charariaw

Distribution : Whole of N.E. Region

Description : A erect shrub 1 to 2 m tall. Leaves digitately trifoliate. Flowers in raceme, purple. Pods oblong, tomentose two seeded.

Uses: A principle source of dye. Pods are eaten. Roots used for external application for ulcers and swelling. Also, used in blindness, cholera, fever fistula, pain in muscles, small pox, spleen complications and stomachache plants grown in gardens as ornamental plants. Host for lac insects.

Flemengia prostrata Fabaceae

Common name : N.A

Distribution : Meghalaya upto 2000 M

Description : A sub-erect or trailing shrub about 1 M tall. Stem angled grooved. Leaves trifoliate. Leaflets narrow lanceolate, cuspidate-acute, base of terminal leaflets oblique and rounded on the outside, thinly pubescent or glabrous above. Racemes very dense. Flowers pink and yellowish green.

Uses: A purple dye is obtained from the glands at the base of hairs covering the pods.

Flemingia semialata Roxb. Ex Ait Fabaceae

Common name: Salpan, Bhatte, A chakmi chong, Papai, Kelbe.

Distribution : Throughout N.E. Region upto a height of 2000 m.

Description : A shrub much taller, leaflets 3-4 nerved at the base. Racemes much longer and lax.

Uses : Vermifuge, bodyache and epilepsy.

Flemingia stricta Roxb. Fabaceae

Common name : Dangshu-khop, Mi-surung,-pu-arong, Kombat-lai-jiri

Distribution : All over N.E. Region in damp localities and ravines in hills.

Description : A sparingly branched shrub, branches, 3-queterous. Leaves digitately trifoliate, lanceolate, acuminate, finly silky outside. Leaflets lanceolate or oblong-lanceolate, gradually narrowed to sharp apex. Raceme spicate dense flowered. When young the inflorescence gives a cone like appearance in bud. Flowers purple.

Uses: As vermifuge and in skin diseases

Flemingia strobilifera (L.) R & Br. Fabaceae

Common name : Makhioti, Makhloti, Ka-Khong

Distribution : Found in grasslands in upper regions of N.E. Region upto 1500m

Description : A shrub usually 2 m tall sometimes as tall as 3 to 4m. Stem blackish. Leaves unifoliate oblong to ovate, lanceolate acute or acuminate rounded at the base, chartaceous, glabrescent above pubescent below. Flowers purplish in 3-4 flowered cymes hidden in membranous bracts.

Uses: Roots used in epilepsy and hysteria. Also induced to reduce pain and induce sleep. Leaves used as vermifuge and also in skin diseases.

Indigo artopurpurea B. & H. Fabaceae

Common name : N.A

Distribution : Meghalaya upto 1500 m

Description : A large shrub with twiggy branches. Leaflets opposite 11-17, oblong obovate, Obvate-oblong, oblong, obtuse, mucronate, Green above glabrescent or paler beneath. Racemes dense towards apex, pedicle short. flowers deep pink or dark purplish.

Uses: Twig is used for basket making and rope bridges.

Indigofera cassiodes Rotle ex DC. Syn. **I. elliptica, I. pulchella** Fabaceae

*Common name:*Eaterjata, Ulhel, Hottar, Nerdi, Sakena

Distribution : Entire N.E. Region.

Description : Shrub, branches striate, alternate. Leaves imparipinnate, leaflets opposite oblong, elliptic, glabrescent. Flowers, bracteate, pale red.

Uses: Epilepsy, labour pain. It is edible and used as vegetable.

Indigo suffruticosa Mill Fabaceae

Common name : Vilayatimal, Shimmali.

Distribution : Assam.

Description : A low undershrub 1 m tall. Leaves long petioled, leaflets 5-15. opposite, obvate oblong, glabrous above, grey hairy beneath. Raceme long peduncled 10-20 flowered. Flowers pale orange.

Uses: Febrifuge, vulnary, stomachic and diuretic. Also used in syphyllis and epilepsy.

Indigo sumatrana Gaertn Syn **I. tinctoria** L. Fabaceae

Common name: Indigo, Neel, Jarmooli, Kiron, Tarhalim

Distribution : Widely cultivated in the region.

Description: Small shrub, branches angular. Leaves minute, subulate, leaflets obovate glabrous above silvery pubescent beneath. Racemes erect. Spike like flowers greenish yellow small on short pedicel.

Uses: Decoction of root and seed is used in urinary diseases and ulcers.

Indigofera tinctoria L Syn. **I. sumatrana** Gaertn Fabaceae

Common name : Indian indigo, Nil

Distribution : Widely cultivated in the region

Description : Shrub about 2 m tall, branches angular. Leaflets glabrous above silvery pubescent beneath. Racemes erect. Flowers small on short pedicel. Flowers greenish yellow wings pink.

Uses: Used as green manure. Grown for extracting indigo dye. Extract used in epilepsy and other nerves disorders. Leaves also used in the form of ointment for sores and old ulcers and piles. Decoction of leaves given in haemorrhage. Root is used in urinary diseases and hepatitis.

Lablab purpureus L. Syn. **Dolichus purpureus, D. Lablab** Fabaceae

Common name : Sem, chicaros, chink

Distribution : Throughout N.E. Region

Description : Annual twining herb. Leaves trifoliate, leaflets ovate acute, base cuneate or deltoid, slightly pubescent beneath. Racemes lax on peduncles. Flowers purple. Stamens diadelphous 9+1.

Uses: Cultivated, pods used as vegetable.

Lathyrus aphaca L Fabaceae

Common name : Jangli matar

Distribution : All over the region

Description : Annual with slender much branched stem, stipules in pairs, foliaceous. Leaves modified to tendrils. Flowers yellow on axillary peduncle.

Uses: Used as fodder.

Lathyrus odoratus L Fabaceae

Common name : Sweet pea

Distribution : Cultivated as an ornamental all over the region in gardens.

Description : Annual climbing herb, glabrescent. Stem winged. Leaflets in pairs, ovate oblong glabrous above hairy beneath, tendril branching. Flowers of various colours.

Uses: Cultivated in gardens as an ornamental.

Lathyrus sativus L. Fabaceae

Common name : Khesari, tewara

Distribution : A useful pulse cultivated in Bhramputra valley upto Darrang

Description : An annual sub-scandent herb. Leaflets lanceolate, entire, auricled. Corolla bright blue or white.

Uses: Used as a pulse crop but sometimes has a toxic substance

Lens culnaris L. Fabaceae

Common name : Masoor

Distribution : Cultivated as a pulse in some parts of N.E. Region

Description : Annual herb, branches angular, Leaflets 4 -6 pairs pubescent lanceolate. Short axillary peduncles bears 2 small white or pale blue flowers and ends in a bristle.

Uses: Seeds used as pulse, Plant used as fodder

Lespedeza juncea (L.f.) Pers. Fabaceae

Common name : N.A.

Distribution : Mosly in North Manipur

Description : An erect undershrub with tough long branches. Leaves trifoliate, ascending crowded, leaflets small linear cuneate, usually glabrous above silky beneath. Flowers 2-4 pale yellow or white tinged with purple borne on short pedicels

Uses: Contains tannins, used for silage and hay. Stock left after harvesting may be used for paper pulp suitable for cheap paper. Seeds produce a semi-drying, oil.

Leucaena leucocephala (Lamk.) Mimosaceae

Common name : Subabool

Distribution : Planted in fields and gardens to supply fuel wood.

Description: A small tree, leaves large bipinnate, pinnae 4-8 pairs. Leaflets 10-15 pairs. Flowers white, sessile in dense globose heads. Stamens 10 free twice as long as corolla

Uses: Tender fruits and seeds are edible. Leaves used as fodder. Bark has medicinal properties. It is good source of green manure.

Lupinus perennis L Fabaceae

Common name : Turmur

Distribution: All over N.E. Region particularly Tripura.

Description : Herbs about 1 m long, pubescent. Leaflets oblanceolate, obtuse, mucronate. Flowers bluish to pink or white on loose racemes

Uses: Cultivated as an ornamental plant in gardens

Macrotyloma uniflorum Lamk. Vevideck Fabaceae

Common name : Kulthi, Patal luri

Distribution : Assam

Description : An annual suberect or twining plant with hairy stem. Leaflets long ovate, acute not lobed. Flowers pale yellow. Calyx downy teeth elongate exceeding corolla tube

Uses: Cultivated in domestic compounds, Regarded medicine for kidney stones.

Mezoneuron enneaphyllum W. & A. Caesalpiniaceae

Common name : Kachai

Distribution : All over N.E. Region

Description : Bushy shrub armed with sharp hooked prickles. Leaf rachis prickly. Leaflets oblong, obtuse, base unequal, glabrous or so. Racemes panicled. Flowers bright yellow.

Uses : Seeds given to cattle as vermifuge

Mezoneuron cucullatum Weight & Arn Caesalpiniaceae

Common name : Bhagsora, Ling-khang, Bango-tatake, Duff-Kuchai-lot

Distrinution All over N.E. Region upto a height 500-1000 m.

Description: A large straggling shrub. Branches glabrous covered with blackish prickles. Bark thin yellowish grey outside red inside Leaf rachis and its branches armed with hooked prickles, leaflets 3-6 ovate, elliptic-oblong, obvate-oblong, bluntly-acuminate base cuneate or rounded. Racemes terminal and axillary often from old wood. Flowers bright yellow. Stamens 5. one exceeding the others 5 short, filaments with few hairs near the base anthers red.

Uses: Seeds given to cows as vermifuge

Milletia extensa Benth and Baker Syn. **M. auriculata** Baker ex Brandis Fabaceae

Common name : Gaunj, Agya, Aqntmala, Makrisiari

Distribution : Assam

Description : A sub-erect shrub more often a woody climber up to 20 m in length. Bark yellowish brown with small rough lenticels. Leaves imparipinnate leaflets oblong, obvate or elliptic. Flowers in axillary racemes whitish or brownish

Uses : The leaves and twigs are used as cattle fodder. The leaves are fairly rich in protein, calcium and phosphorus. Root is reported to possess excellent repelling properties. Used as fish poison and also applied to sores of cattle and horses to keep of flies. Bark and stem yield coarse fibre used as cordage. they are beaten and used as brushes

Milletia fruticosa Benth and Baker Fabaceae

Common name : N.A.

Distribution : All over N.E. Region particularly Tripura

Description : Closely resembles *M. auriculata* Benth and Baker and has numerous leaflets ranging between 9-11, Subsessile. Flowers reddish in shorter axillary racemes and also in terminal panicles and nearly glabrous pods

Milletia pachycarpa Benth Fabaceae

Common name : Bokao-bih, Holosi,, Mak-Khal, Ru- go-jao, Arru, Kuario

Distribution : Meghalaya upto 1500 m

Description : A large climber, branches terete. Leaves decidous, leaflets exstipulate 11-13, terminal obvate, lateral oblong to lanceolate to oblanceolate chartaceous, abruptly acuminate. Racemes somewhat densely flowered axillary or terminal panicles. Flowers in fasicles at the nodes, sometimes from short peduncle. Flowers light mauve. Stamens monoadelphus

Uses: The roots and seeds used as poison for fish and birds. Also possess insecticidal activity. Toxicity due to presence of Rotenon and related compounds extract and aqueous suspension of ground seeds acts both as stomach and contact poison, also ovicidal. Seeds contain oil

Milletia caudate Baker Fabaceae

Common name : N.A

Distribution : Assam Meghalaya, Manipur upto 1000 m

Description: An erect or straggling shrub about 5 m tall branches sparsely lenticelled. Leaves glabrous. Leaflets oblong-elliptic to oblong-lanceolate, bright green above glaucous beneath. Racemes axillary lax flowered. Flowers reddish, diadelphous.

Milletia pulchera (Benth.) Kurz Fabaceae

Common name : Dieng-Shakuriao, Dieng-tiw-khmat. TAW-tynnery, So-long-tew

Distribution : Meghalaya from 1500 –2000-m

Description : An erect shrub or small tree 8-10 m tall. Young parts densely rusty, branchlets bark thin, dark grey dull white outside. Leaves with rachis, dull pubescent. Leaflets 15-21, narrow, elliptic oblong and bluish grey beneath. Racemes axillary, simple. Flowers purple fascicled on slender pedicel.

Uses: The wood is very tough and fit for tool handles and agricultural implements

Mimosa himalyana Gamble Mimosacea

Common name : Kuchoi, Kaitkusia-kanta, Ingsu-maha, Jharua, Kalar sarkhi

Distribution : Assam, Meghalaya, often in clearances and old jhum

Description : A large shrub, bark smooth and reddish brown on young branches. Leaves bipinnate, leaflets 10-20 pairs on each pinnae lowest pair often reduced to bristle, oblique oblong glabrous and dark green above, paler and sparingly hairy beneath. Heads which are either fasicled on leaf axils towards the end of the branches or arranged in terminal panicles. Flowers pink or white. Ovary glabrous, stipitate, many ovuled, Style filiform, Stigma terminal

Uses : Wood useful for tent pegs and gun powder charcoal

Mimosa pudica L. Mimosaceae

Common name : Varah-Kranti, lajjalu, lajjabati, sharmida, nilja-bon. Lajula-pata, saji-gablau, Lajura, jharuachickanta, samukchip

Distribution : Common in wasteland throughout N. East Region

Description: A spreading, struggling,decidous under-shrub. Stem and branches prickly and clothed with bristles. Leaves sensitive, pinnae 4, digitately arranged leaflets 12-20 pairs, obliquely narrowed, oblong, acute. Flowers purple, Stamens 4, much exserted.

Uses: The roots are used medicinally. Juice of leaf used for dressing ulcers, sores and piles. Seeds yield a fatty oil similar to soyabean oil.

Mucuna bracteata DC. Fabaceae

Common name : Wakami

Distribution : Common in region and grown upto 1000 m.

Description : An annual or perrenial climber, leaflets, terminal rhomboid lateral deltoidly oblique, acute or sub-acute, dark green, glabrescent above pubescent beneath. Racemes axillary bracteate down to the base. Flowers blackish purple. Pods covered with reddish bristles which are irritating.

Uses: The seeds are used as an aphrodiasc and as vermifuge. A delicacy in Meghalaya.

Mucuna cochinensis Cheval Syn. **M. nivea** Fabaceae

Common name : Kahwanch, Kursi

Distribution : All over N.E.Region

Description : An annual twining herb with white or pale purple flowers and glabrescent pod. Pods downy when young, seeds 6-8, oval, smooth. Ash coloured.

Uses: Cultivated for its edible pods. Useful as fodder, cover crop and green manure.

Mucuna deeringiana (Bort.) Merrill Fabaceae

Common name : Makhmali sem

Distribution : All over N.E. Region

Description : A herbaceous pubescent trailing and twining annual herb few meters in length. Leaves trifoliate, leaflets ovate, rhomboid. Flowers purple in pendant raceme, seeds 3-5 in a pod brown or black

Uses: Used as green manure, cover crop and also as fodder. Young pods edible

Mucuna monosperma DC ex Wight Fabaceae

Common name : Mei-sia-ryntim

Distribution : Swampy localities and foot hills of the mountains.

Description : A large woody climber branches longitudinally ribbed. Leaves ferruginous-pubescent. Leaflets ovate oblong the lateral very oblique abruptly acuminate glabrous above pubescent beneath. Racemes axillary from old nodes. Flowers with sub-orbicular standard and dull purple in colour. Pod ovate ovoid 1 seeded densely covered with dark brown irritating bristles

Uses: Seeds are restorative. and posses expectorant properties and is used in cough asthama and affectations of tounge. They are said to be sedative. The trichomes of pods are irritant to skin and may cause dermatitis. Young seeds used as vegetable

Mucuna nigricans (Lour.) Steud Fabaceae

Common name : Mikuri-ghilla

Distribution : Assam

Description : A large woody twiner with dull purple flowers and oblong pods clothed With yellowish brown irritating bristles.

Uses: Watery sap from the stem used in cough and fever. Pod hairs produce dermatitis.

Mucuna puriens (L.) DC. Syn **M. purriata** Fabaceae

Common name : velvet bean

Distribution : All over N.E. Region

Description : Climber with adpressed pubescent branches, leaflets ovate-rhomboid, glabrescent above grey silky beneath. Flowers purple in short drooping peduncled racemes

Uses: Fruits are used as vegetable. Roots and fruits are medicinal.

Neptunia olearaceae Lour Syn. **N. prostrata** Mimosaceae

Common name : Lagalu, Pani Nagak

Distribution : In swamps allover N.E. Region

Description : An aquatic herb, with widely creeping or floating stem rooting at the nodes and producing sub-erect branches. Leaves bipinnate pinnae 4-6, leaflets sensitive 5-15 pairs oblong, obtuse, glabrous.Flowers minute in dense heads the lower replaced by numerous ligulate staminodes Stamens exerted.

Uses: Used as vegetable. Juice of stem used in earache.

Oromosia assamica Yakovlev Fabaceae

Common name : Porghum, Sanachi, Sarachi-bloma

Distribution : Meghalaya, Assam

Description : A lofty tree upto 30 m tall Bark greenish grey or brown. Leaves imparipinnate leaflets 7-11, opposite elliptic oblong or lanceolate or oblong-lanceolate usually bluntly acuminate. Flowers dull white in terminal panicles the lower branches of which are from the axils of upper leaves. Stamens 10, free

Uses: Wood light brown but fairly tough used locally for domestic purposes

Parkia timoriana (DC.) Merrill Syn. **P. javanica, P. roxburghii** Mimosaceae

Common name : Yongchak, Lonchak. Barri-phong

Distribution : All over the N.E. Region

Description : A middle sized tree with spreading branches. Leaves bipinnate main rachis puberolous or glabrescent with a gland below the lower pair of pinnae. Pinnae 8-30 pairs opposite. Leaflets 40-80 pairs, sessile, oblong falcate truncate at base. Flowers yellow sterile in dense clavate long peduncled heads.

Uses: Tender pods are eaten and also used for flavoring. Seeds though slightly bitter are eaten after roasting. Pods and seeds are used for stomach disorders. Decoction of bark and leaves applied to sores and skin affections.

Parochetus communis Hamilt B. & H. ex. D.Don Fabaceae

Common name : Khia knoll.

Distribution : All over the region trailing along the road sides and forests upto 1500m.

Description : A slender creeping herb with tufted runners. Leaves digitatedly trifoliate leaflets 2 sub sessile broadly obvate-obtriangular or obcordate slightly emargi-nate at the apex, mucronate, green and glabrous above gre-silky beneath. Flower blue with reddish veins. Stamens dia delphous, Style elongated.

Uses: Has medicinal properties. Used in stomach disorders in children.

Peltophorum inerne (Roxb.) Syn. **P. pterocarpum** Caesalpiniaceae

Common name : Copper pod

Distribution : Tripura and adjoining states. Grown as shade tree

Description : Large evergreen trees, bark smooth, grey. Leaves bipinnate 16-20 leaflets in 10-15 pairs obliquely oblong, obtuse. Flowers yellow in large erect terminal panicled racemes rusty tomentose. Stamens 10 unequal with orange anthers

Uses : Cultivated as shade tree. Bark used for extracting tannin. Used in dysentery for preparing gargles, tooth powders and lotions used in eye troubles. Leaves used as cattle feed.

Piptanthus nepalensis (Hook.) D.Don Fabaceae

Common name : N.A.

Distribution : Throughout N.E. Region particularly in Meghalaya

Description : A shrub 3-4 m tall somewhat gregarious in habit branches with verticle ridges and annulate with scars of fallen stipules. Leaves digitatedly trifoliate, leaflets sessile lanceolate or narrowly elliptic acuminate, entire, dark green above pale beneath Flowers yellow in short compact hairy racemes. Stamens 10, free ovary linear hairy

Pithecellobium dulce (Roxb.) Benth Mimosaceae

Common name : Jungle jalebi, Kachlora, Vilayati imli

Distribution : Found all over N.E. Region

Description : Large ever green tree, Bark grey, branchlets armed with spines. Leaves bipinnate, pinnae 2, each with a pair of leaflets. Flowers white sessile in small globose heads arranged in a long narrow terminal panicle. Stamens many monoadelphous

Uses: Grown as avenue tree. Seeds edible.eaten raw or in curries. Saline extract of seed shows haemolytic agglutinating reaction with human blood. Yields a fatty oil used as a cooking medium and also in soap making

Pithecelobium heterophyllum (Roxb.) Macb Mimosaceae

Common name : Arda, Morangmara

Distribution : Tripura an other adjoining States

Description : Middle sized tree with grayish bark. Leaves bipinnate, pinnae 2-4 pairs with a gland just below the insertion of leaflets, leaflets 2-10 pairs, oblong to lanceolate to elliptic or rhomboid-ovate dark green and glabrescent above. Flowers yellowish white in peduncled umbellate corymbs

Uses: Fruit gives a black dye.

Pithecelobium monodelphum Bedd. Mimosaceae

Common name : Bhachahu, Dieng-yep-yai

Distribution : Assam and Meghalaya upto 6000 m

Description : A small to medium sized tree. Bark smooth thin brown. Leaves bipinnate pinnae 1-2, pinnules elliptic-lanceolate or ovate or oblong. Flower heads yellowish white

Uses: Bark contains a chemical pithocellobin which is used as fish poison. Seeds also contain it and cause vomiting if eaten uncooked or without roasting. A decoction of leaves is used as an external application in leprosy. Leaves also contain saponin and an essential oil.

Pongamia pinnata (L.) Pierre Fabaceae

Common name : Karanja, Pitagaria

Distribution : Tripura and other adjpoining states. Cultivated as an ornamental plant

Description : A small tree with spreading crown, bark gray. Leaves imparipinnate, leaflets opposite, ovate elliptic, acuminate obtuse at base, glabrous dark green above pale beneath. Flowers white or purple in simple axillary or peduncled racemes. Stamens monodelphous

Uses: Fruit is edible. Leaves are used as green manure. Seed oil is medicinal and also used in manufacture of soap.

Priotropis cytisoides W & A Fabaceae

Common name : Dieng-sha-ratoh, Chakuriao, Dieng-toh-tori

Distribution : All over N. E. Region up to 2000 m

Description : A shrub somewhat bushy in habit. Leaves long slender slightly channeled leaflets 3, oblong elliptic to lanceolate acute at both ends. Racemes terminal. Flowers pale yellow standard and keel turning purplish. Stamens monodelphous.

Uses: Used as green manure for paddy

Prospis cineraria (L.) Bruce **P. specigera** Mimosaceae

Common name : chonker, khejri, sanger, Shami

Distribution :: Not very common in the area

Description : Middle sized tree armed with straight prickles. Leaves bipinnate pinnae 4 opposite with a gland between each pair. Leaflets 8-12 pairs. Flowers small yellow in slender spikes arranged in short peduncled axillary panicles. Stamens 10, free.

Uses: Grown as hedge plant. Pods, bark and flowers are medicinal as a matter of fact every part of the plant is used. Leaves used as fodder and fruits as vegetable after drying.

Psophocarpus tetragonolobus DC. Fabaceae

Common name : Chara koni sem, Lakar sem, Goa bean

Distribution : All over N.E. Region

Description : A slender climber with large tuberous roots. Leaves trifoliate. Leaflets, ovate acute deltoid at base. Flowers rather large lilac in lax racemes Stamens monodelphous.

Uses: Roots and fruits are used as vegetable. As a matter of fact every part of the plant is used.

Pureria phaseoloides (Roxb.) Benth. Fabaceae

Common name : Jermei-kyn-saw. Jermei-soh-gousoh

Distribution : Whole of N.E. Region

Description : A twining annual often trailing on sandy beds of streams. Young shoots covered with patent brown hairs. Leaflets broad, sub-rhomboid to broad ovate lateral oblique, obtuse, acute very shallowly lobed or sometimes not lobed. Racemes simple long peduncled. Corolla pale blue or reddish stamens generally free. Pod black covered with stiff adpressed hairs.

Uses: A useful strong white fibre is extracted from the stem which is used for making twins and ropes.

Pureria thunbergiana Benth Fabaceae

Common name : Suting, Suting-rit

Distribution : Common in hilly areas of N.E.Region upto 1500 m

Description : A large climbing shrub, branches slender clothed with fine deflexed brownish hairs. Leaflets rhomboid to ovate lateral oblique sharply acuminate, glabrous above and thinly covered with very short adpressed hairs beneath Racemes simple densely flowered towards the extremity. Flowers purple or reddish, sweet scented. Stamens free.

Uses: Used as cattle fodder.

Purerария tuberosa Roxb. ex (Wlld.) DC. Fabaceae

Common name : Bilai kand,Ghorabel, Patal kumra, Pithania, Vidarkand.

Distribution : Common in sub-himalayan tract of the region particularly Assam.

Description : A large woody climber with very large tuberous roots. Leaflets terminal broad, ovate, or rhomboid, silky pubescent when young, glabrous above and thinly adpressed pubescent beneath. Racemes erect, usually panicled and terminal or from axils of the fallen leaves. Flowers bright blue fading to bluish purple. Stamens usually diadelphous. Seeds with long brisle like brown silky hair.

Uses: The tubers have a liqorice like taste and are eaten raw. Also used in medicine.

Pycnospora lutescence (Poir.) Schindl. Fabaceae

Common name : Dariavel

Distribution : All over N.E. Region

Description : A perennial plant. Leaves pinnately trifoliate, petiole flattened sparsely ciliate. Flowers small in slender axillary racemes. Flowers purplish stamens diadelphous 9+1. Ovary sessile many ovuled.

Uses: Used as fodder and green manure

Rhynochosia rothii Benth ex Ait Syn. **R. sericea** L Fabaceae

Common name : Bhim- khatrenj

Distribution : All over N.E. Region

Description : A woody spreading climber, young parts, glandular-pubescent. Leaves usually trifoliate, leaflets broadly ovate, acute, terminal one somewhat deltoid. Racemes lax long peduncled. Flowers yellow

Uses: used as abortif and also as cure for asthama and piles. Also as green manure.

Rhynochosia rufescence (Willd.) DC. Fabaceae

Common name : N.A.

Distribution : All over N.E.Region

Description : Climbers with slender branches. Leaves pinnately trifoliate ovate or rhomboid acute sub-coriaceous. Racemes 2-6 flowered. Flowers yellow.

Uses: As green manure and used as medicine in asthama.

Rhynochosia scarabaeoides DC Fabaceae

Common name : Ban kalai

Distribution : Throughout N. E. Region

Description: Profusely branching herb, branches slender twining. Leaflets terminal, elliptic or obvate, oblong. Flowers yellow axillary or few in racemes. Stamens diadelphous 9+ 1 sessile.

Uses: Used as green manure and as a medicine in asthama.

Samanea samen (Jacq.) Merril Mimosaceae

Common name : N.A.

Distribution : All over N. E. Region

Description : Large decidous tree with spreading crown. Leaves bipinnate, pinnae 3-7 pairs with 2-10 pairs, leaflets opposite, ovate, oblong, oblique shining above. Flowers pink in solitary globose heads on pubescent peduncles from the upper axils. Stamens 20, pink connate below.

Uses: Planted as avenue tree. Leaves and pods used as fodder. Tree yields a gum which swells in water into a tough cartilage like mass.

Saraca asoca (Roxb.) de Wide Caesalpiniaceae

Common name : Ashok

Distribution : All over the region planted as ornamental plant.

Description : A tree with umbreageous (dark) crown, bark blackish, wood reddish. Leaves paripinnate, leaflets 4-6 pairs large, oblong, coriaceous. Flowers fragrant, orange-scarlet in compact panicles of corymbs from leaf axil or nodes. Stamens 6-8 much exerted, anthers purple.

Uses: Grown as an ornamental plant, Bark has medicinal properties.

Sesbania bispinosa (Jacq). W.F. Wight Fabaceae

Common name : Canicha, Danchi, Dahencha,

Distribution : Assam, Tripura and other plain areas

Description : A large annual erect shrub. Branches and underside of leaf rachis armed with small hooked prickles, leaflets numerous. Flowers rather small and yellow.

Uses : Used as green manure. Stems provide a strong durable fibre substituted for hemp in rope, cordage for fish net. Seeds are fried and after mixing with wheat flour applied to ringworm and other skin diseases.

Sesbania cannabina Retz Poir Syn. **S. aculeate** Baker Fabaceae

Common name : Jayanti, Ilkata, Dhunchia

Distribution : Assam, Tripura

Description : An annual shrub, branches slender obscurely muricated. Leaves imparipinnate leaflets numerous linear oblong. Racemes sub-sessile 3-4 flowered

Uses: Used as green manure and as fodder. Also grown for fibre for sails fishing net lines and ropes. Seeds yield a gum similar to guar gum. Plant useful in saline and alkaline soils

Sesbania grandiflora (L.) Poir Syn. **S. formosa** Fabaceae

Common name : Bak phul, Agati

Distribution : All over N.E. Region

Description : A soft wooded tree bark light brown, wood white, Leaflets 40-60 linear oblong, glaucous. Flowers large, axillary in few flowered racemes, white or reddish showy, Stamens diadelphous 9+1

Uses: Flowers and immature pods are used as vegetable. Leaves used as fodder and green manure. Bark yields a fibre

Sesbania sesban (L.) Merrill Fabaceae

Common name : Jentis, joyontri

Distribution : Assam

Description : A glabrous unarmed soft wooded shrub or a small tree with slender terete branches. Flowers yellow in lax axillary racemes.

Uses : Cultivated as an ornamental plant. Bark fibre is used as a substitute for hemp for making ropes. Bark, leaves and seeds are medicinal and used in headache and mumps. Young pods edible

Shuteria vestita W & A Syn. **S. involucrate** (Wall) Wight & Arn

Common name : N.A.

Distribution : All over N. E. Region in forest glades upto 1500 m

Description : A very slender climber with spreading hairs on the stem. Leaves trifoliate very slender patently hairy, leaflets ovate obtuse subcordate at base glabre-scent above pubescent beneath. Raceme sessile dense often fasicl-ed. Flowers purple. Stamens diadelphous. Ovary many ovuled

Uses: As green manure in tea and cinchona plantation

Smithia blanda Wall ex Wight & Arn Syn. **S. racemosa** Fabaceae

Common name : Bat-shakuriao, Barka

Distribution : Common in water logged places and edges of swamps upto 1500 m.

Description: A perennial suffruticose herb Stems clothed with fine yellow bristles leaflets 6-10 subsessile oblong cuneate or rounded at the base membrabeous but firm with bristles along the edges and midrib. Racemes panicle terminal. Flowers bright yellow. Pods dark brown reticulated.

Uses : Used as vegetable.

Smithia sensitivia Ait Fabaceae

Common name : Munga jori

Distribution : In grassland in hillocks upto a height of 1000 m

Description : A much branched semi-woody weedy plant about 1m tall. Leaflets generally 4-12, linear ciliated with bristles mainly along the lower midrib. Racemes axillary simple peduncled, pedicels short. Flower bright yellow with a brown band in the throat.

Uses: Leaves eaten as pot herb. Makes excellent hay. Herb boiled and given in difficult micturation.

Tamarindus indica L Caesalpiniaceae

Common name : Tntiri, Amlika, Tentul, Tetuli. Imli. Khen tikiri, Thei-apai, Dieng soh tantai

Distribution : Throughout N.E. Region

Description : A large handsome evergreen tree with shallow fissured, dark grey rough bark. Leaves pinnate leaflets 10-20 pairs. Linear oblong. Racemes lax, terminal, few flowered. Flowers pale yellow or cream coloured. Stamens monodelphous, 3 perfect rest reduced to 3-4 setaeous staminodes.

Uses: The tree is grown for pulp. Pod to prepare beverage and flavor curries and sauces. It is antiscorbutic. Pulp is used in dysentery, malaria, jaundice eye diseases and leprosy. It is aperient and febrifuge

Tephrosia candida (Roxb.) DC. Fabaceae

Common name : Bilakshani, Bilokheri, Samendu, Bolmendu, Balasor, Bangora, Dien-toh, Nathri, Mith-phang

Distribution : Throughout N. E. Region

Description : A shrub 2to 3 m tall with gregarious tendencies branches angled or grooved. Leaflets 15-30, narrowly oblong to obvate-oblong or ligulate, minutely mucronate sub-glabrous above pale and densely adpressed silky beneath base acute. Racemes simple, terminal often with a few shorter ones from the axils of the upper leaves. Flowers creamy white in fasicles of 2-4 styles flattened,silky

Uses : It is used to improve the soil conditions in new area are as a nursery for seedlings in forest plantations. Bark and leaves used as fish poison. Plant has been reported to have medicinal properties.

Tephrosia purpurea (L.) Pers Fabaceae

Common name : Ban-nil, Sarapunkha

Distribution : All over N.E. Region

Description : A sub-erect diffuse perennial small shrub, branclets terete, often glabrescent often more or less zigzag, young parts silky. Leaflets 7-21, obovate, rounded or emarginated, shortly mucronate, cuneate at the base, glabrescent above and hairy beneath. Raceme lax flowered often leaf opposed. Flower purple. style flattened

Uses: Tonic, laxative, diuretic. Used in bronchitis, boils, pimples and bleeding piles decoction of root is given in diorrhea, rheumatism,asthama and urinary disorders

Uraria clarkei Gagnap Syn. **U. lacel craib** Fabaceae

Common name : Mirong-rondai.

Distribution : In hills where plants have been cut down or in jhum lands.

Description : A shrub about 1.5 m tall with ferruginous, hairy more or less sulcate stems. Leaves pinnately trifoliate, terminal leaflet largest, ovate, lanceolate, obtuse mucronate at the apex, adpressed with pubscent with somewhat harsh hairs above, pilose beneath specially on the nerves. Flowers in terminal ferruginous panicles.

Uses : Used in ringworm and in sores and is also known to be antipyretic

Uraria crinita (L.) Desv. Fabaceae

Common name : Dieng-kha-riu

Distribution: Fairly common in Meghalaya upto 1500 m

Description : A shrub 1-2 m tall, Stems stout strigose with hooked hairs when young. Leaflets 3-7, oblong, acute or sub-acute at the apex, minutely mucronate generally rounded at the base, coriaceous, smooth, glabrous and pubescent along nerveules beneath. Racemes terminal, very hairy, shaggy with spreading brown bristles. Flower purple delicately veined

Uses: Used as fish poison. Leaf decoction kills lice. Shrub is also reported to be used in diorrhea, dysentery, enlarged liver and spleen also in pustules and tumors.

Uraria lagopoides (L.) Desv Fabaceae

Common name : Ghursai, Ishargata, Mahadelijata

Distribution : All over N.E.Region upto 1500 m

Description : A little branched erect under shrub. Stems terete. Leaves very variable leaflets sub-coriaceous, glabrous and variegated with irregular greyish branches. Racemes terminal, simple very dense and hairy. Flowers purple.

Uses: Roots and pods used in prepration of ringworm ointment. Decoction of leaves is used in diorrhea and also as abortifacient. Used as green manure also.

Uraria rufescens (DC) Schindl Syn. **U. hamosa** Wall ex Wright Fabaceae

Common name : Prioniparni, Pithavam

Distribution : In grass ravines in sal areas of N.E. Region

Description : A diffuse undershrub. Young parts clothed with long hairs and shorter hooked bristles. Leaves trifoliate, leaflets oblong, elliptic or ovate, lanceolate, glabrous above downy below, Racemes copiously panicled.

Uses: The decoction of the leaves in combination with some other drugs used as febrifuge

Vicia faba (L.) Fabaceae

Common name : Horse bean, Windsor bean,Trek bean, Fava bean, Broad bean

Distribution : All over N.E. Region

Description : An Annual herb, Stems large, unbranched with one or two stems from the base. Leaves compound, leaflets usually 6 large broad oval. Flowers white with dark purple marking in clusters of 1-5 in axils of leaves,1-4 pods developing from each flower.

Uses: Cultivated as vegetable both green and dried. Used as stock feed. Nutritional value very high. Most important winter garden vegetable. Also used as green manure.

Vicia hirsute S.F Grey Fabaceae

Common name : Jhunjhuni

Distribution : Tripura and adjoining states

Description : A prostrate annual with pubescent stem. Leaves paripinnate ending in twisted tendrils, leaflets 5-8 pairs. Flowers in peduncled axillary racemes, 3-6 flowered. Flowers pink or purple. Stamens diadelphous.

Uses : Seeds contain hydrocynic acid.

Vicia sativa L Fabaceae

Common name : Common vetch., Akrg. Ankarg

Distribution : Uncommon in North East. Found in patches in drier parts

Description : A struggling or ascending robust annual herb, stem 10 –70 cm long, glabrescent to sparsely pubescent with short spreading hairs. Leaves compond with 4-12 leaflets, variable in shape, linear, narrowly oblong to broadly obvate or obcordate upper leaves with well developed branched tendril. Flowers solitary or paired, subsessile light brown to straw coloured.

Uses: Grown in area as with mild winter crop largely grown for seeds. Seeds have a chemical which causes problems to nerves. Seeds are said to be astringent and a remedy for diorrhea.

Vigna pilosa (Willd) Fabaceae

Common name : Jhikrai, milkonia

Distribution : Tripura and other adjoining states

Description : A climber, branches very slender. Leaflets ovate acute with short adpressed hairs. Racemes many flowered. Flowers reddish. Stigma oblique with a tuft of hairs below the tip of style.

Uses : Straw fed to cattle and grain eaten in time of scarcity

Vigna umbellata (Thunb) Ohwi and ohashi Fabaceae

Common name : Rice bean, climbing bean, Mambi bean, Oriental bean

Distribution : Allover N. Eastern Region

Description : Annual herb, stem sub-erect. 30 to 75 cm tall producing twining branches. Leaves trifoliate. Leaflets ovate. Inflorescence erect, axillary racemes with 5-20 flowers, bright yellow.

Uses: Widely grown and used as pulse. Young pods and leaves are used as vegetable whole plant used as fodder for cattle and also as cover crop and as green manure.

Vigna unguiculata (L) Walp Fabaceae

Common name : Barbati, Cowpea

Distribution : All over N.E. Region grown as vegetable in the kitchen garden

Description : A twining annual with glabrous stem, Leaves trifoliate, leaflets entire or slightly lobed, ovate lateral ones also lobed, ovate. Flowers in sub-capitate racemes. Flowers yellow. Pods 10-60.

Uses: Fruits as vegetable and dried seeds as pulse

Vigna vexillata (L.) A. Rich Fabaceae

Common name : Jemeim Soh-long-tor

Distribution : Assam and Meghalaya up to 1000 m

Description : A slender twiner with fusiform tuberous roots. Stem terete more or less scabrid with deflexed hairs. Leaves trifoliate, leaflets narrowly lanceolate, finely acuminate more or less rounded or truncate at the base. Flowers so congested at the end of peduncles as to appear in heads. Flowers purple very finely veined with deeper purple.

Uses: The seeds are used as pulse and tuberous roots as vegetable.

Zorina gibbbosa Span Fabaceae

Common name : Landegu, Jeepinghas, Samrapni

Distribution : All over N. E. Region

Description : Diffused annual with slender zigzag wiry branches. Leaves petiolate leaflets in pair, lanceolate, acute, glabrous, dotted with black glands. Flowers yellow in lax axillary racemes. Stamen monodelphous. Stigma minute capitate

Uses : Root decoction is given to children to induce sleep. It is a nutritious fodder for cattle but may cause impairment of eye sight in horses. Also used as cover crop and green manure.

PART - II
Useful Plants of N.E. Region

MEDICINAL PLANTS

1.	*Abrus precatorius* L	Fabaceae
2.	*Achyranthes porphyristachya* Wall	Amaranthaceae
3.	*Aconitum heterophyllum* Wall ex Royle	Ranunculaceae
4.	*Acorus calamus* L	Araceae
5.	*Adhatoda zeylanica* Medik	Acanthaceae
6	*Ageratum conyzoides* L	Asteraceae
7.	*Agrimonia pilosa* Ledeb	Rosaceae
8.	*Allium chinense* Don	Liliaceae
9.	*Allium hookeri* Thew	Liliaceae
10.	*Allium tuberosum* Rottel ex spreng	Liliaceae
11.	*Alocasia macrorrhiza* (L) G. Don/Schott	Araceae
12	*Alnus nepalensis* D. Don	Betulaceae
13,	*Aloe barbadensis* Mill	Liliaceae
14.	*Alpinia galanga* (L) Willd.	Zingiberaceae
15.	*Alstonia scholaris* R.Br.	Apocynaceae
16.	*Alysicarpus monilifer* DC	Fabaceae
17.	*Amomum aromaticum* Roxb.	Zingiberaceae
18.	*Anaphalis adnata* DC	Asteraceae
19.	*Antidesma bunius* Spreng	Euphorbiaceae
20.	*Aphnamixis polystachya* (Wall.)	Meliaceae

21	*Ardisia solanaceae* Roxb	Myrsinaceae
22	*Argemone mexicana* Linn.	Papavaraceae
23.	*Argyreia nervosa* (Burm f.) Boj	Convoluvalacae
24.	*Aristolochia sacata* Wall	Arstilolochiace
25.	*Aristolochia tagala* Cham	Arstilochiaceae
26.	*Asclepias currassavica* Linn.	Asclepiadaceae
27.	*Artemisia nilagirica* C.B. Clarke	Asteraceae
28.	*Asterlbescens* (DC) Koehne	Asteraceae
29.	*Barringtonia acutangula* (L) Gaertn	Lecythidaceae
30.	*Basella alba* Linn.	Baselliaceae
31.	*Bauhinia purpurea* Linn.	Caesalpiniaceae
32.	*Begonia josphi* A. DC	Begoniaceae
33.	*Begonia picta* Sm	Begoniaceae
34.	*Berberis asiatica* Roxb. DC	Berberidaceae
35.	*Bergenia cillata* (Haw) Sternb	Saxifragaceae
36.	*Biophytum sensitivum* DC	Oxalidaceae
37.	*Boehmeria clidemoides*	Urticaceae
38.	*Boenninghausina albiflora* Reicheb ex Meissn	Rutaceae
39.	*Brucea mollis* Wall ex Cruz	Simaroubaceae
40.	*Caladium bicolor* (Ait) Vent	Araceae
41.	*Calamus erectus* Roxb.	Arecaceae
42.	*Calotropis gigentea* (Linn) R.Br	Asclepiadaceae
43.	*Camelia sinensis* O. Kuntz.	Theaceae
44.	*Cavallia brachiata* (Lour) Merr.	Rhizophoraceae
45.	*Carica papaya* L.	Caricaceae
46.	*Cannabis sativa* L.	Cannabinaceae

47.	*Cassia alata* L.	Caesalpiniaceae
48.	*Cassia occidentalis*	Caesalpiniaceae
49.	*Cassia obtusifolia* (L.)	Caesalpiniace
50.	*Casearia veraca* Roxb.	Samydaceae
51.	*Centella asiatica* (L) Urb.	Apiaceae
52.	*Chepanthes farinose* (Kaulf.)	Chenopodiaceae
53.	*Chenopodium ambrosiodes* Linn.	Chenopodiaceae
54.	*Cinnamomum bejalghota* (Buch-Ham.) Sweet	Lauraceae
55	*Cinnamomum tamala* Nees	Lauraceae
56.	*Cissus javanica* Linn.	Vitaceae
57.	*Cissus quadrangularis* Linn.	Vitaceae
58 .	*Cissus repens* Lamk.	Vitaceae
59.	*Clausena excavata* Burm.f	Rutaceae
60.	*Cleisotoma striatum* (Rehb.f) Garey	Orchidaceae
61.	*Cleome viscosa* (L.)	Cappridaceae
62.	*Clerodendron wallichii* Merrill	Verbenaceae
63.	*Clerodendron serratum* Spreng	Verbenaceae
64.	*Coptis teeta* Wall.	Ranunculaceae
65.	*Clinopodium ambrosa* Benth.	Lamiaceae
66.	*Clinogyne stricta* (D. Don) Schl.	Marantaceae
67.	*Coffea bengalansis* Heyene ex Roem & Schultz	Rubiaceae
68.	*Colocasia esculenta* Schott Syn. *C. antiquorum* Schott	Araceae
69.	*Commelina paludosa* Blume	Commelinaceae
70.	*Corydalis longipes* DC.	Fumariaceae
71.	*Coix lacryma* Jobi (L.)	Poaceae

72.	*Costus specious* (Koeing) Sm	Zingiberaceae
73.	*Crassocephalum crepidioides* (Benth) Moore	Asteraceae
74.	*Cotoneaster congestus* Baker	Rosaceae
75.	*Cotoneaster tiglium* L	Rosaceae
76.	*Crotalaria occulata* Grab	Fabaceae
77.	*Cucurbita moshchata* Duch ex Pair	Cucurbitaceae
78.	*Curcuma angustifolia* Roxb	Zingiberaceae
79.	*Curcuma amoda* Roxb.	Zingiberaceae
80.	*Curcuma aromatica* Salisb	Zingiberaceae
81.	*Dalhousiea bracteata* Grah	Fabaceae
82.	*Daphne papyraceae* Wall Syn. *D. cannabina* Wall	Thymelaeaceae
83.	*Datura metel* L	Solanaceae
84.	*Dendrobium nobile* Lindl	Orchidaceae
85.	*Desmodium microphyllum* DC	Fabaceae
86.	*Desmodium-triquetrum* DC	Fabaceae
87.	*Dicranopteris linearis* (Burm.f)	Gleicheniaceae
88.	*Dillenia indica* Linn	Dilleniaceae
89.	*Dipsacus inermis* Wall	Dipsacaceae
90.	*Dioscorea alata* L	Dioscoreaceae
91.	*Diosopyros lanceaefolia* Roxb	Ebenaceae
92.	*Disporum cantoniense* (Lour) Merr	Liliaceae
93.	*Dracaena angustifolia* Roxb	Liliaceae
94.	*Dracaena petiolata* HK.f	Liliaceae
95.	*Drymaria cordata* (L) ex Willd ex R&S	Caryophyllaceae
96.	*Duchesnea indica* Fockes	Rosaceae
97.	*Eclipta prostrata* (L)	Asteraceae

98.	*Elephantopus scaber* Linn	Asteraceae
99.	*Elatostema integrifolium* (D.Don) Wedd	Urticaceae
100.	*Elsholtzia blanda* Benth	Lamiaceae
101.	*Elsholtzia fruticosa* (D.Don) Rehder	Lamiaceae
102.	*Emblica officinalis* Gaertn	Euphorbiaceae
103.	*Entada pursaetha* DC	Mimosaceae
104	*E. phaseeoloides* (L) Merr	Mimosaceae
105.	*Epilobium clarkeanum* Hausskn	Onageraceae
106.	*Erigeron bellidiodes* Benth	Asteraceae
107	*Eryngium foctidum* Linn	Apiaceae
108	*Eupatorium adenophorum* Spreng	Asteraceae
109	*Eupatorium cannabinum* L	Asteraceae
110.	*Euphorbia antiquorum* Linn	Euphorbiaceae
111.	*Euphorbia himalyana* Boiss	Euphorbiaceae
112.	*Euphorbia hirta* L	Euphorbiaceae
113.	*Euphorbia ligularia* Roxb.	Euphorbiaceae
114.	*Euphorbia uniflora* Roxb.	Euphorbiaceae
115.	*Eurya acuminata* DC	Theaceae
116	*Eusetralis stellata* (Lour). Panig	Lamiaceae
117	*Fagopyrum dibotrys* (D. Don.)	Polygonaceae
118.	*Fagopyrum esculentum* Moench	Polygonaceae
119.	*Fimbristylis falcata* Vahal	Cyperaceae
120	*Flemingia macrophylla* (Willd) ex Merrill	Fabaceae
121	*Floscopa scandens* Lour	Commelinaceae
122.	*Fragaria indica* Andr	Rosaceae
123	*Galinsoga quadrirata* Ruiz & Pavan	Asteraceae
124	*Garcina cowa* Roxb.	Guttiferae/

		Clusiaceae
125.	*Gaultheria nummulariodes* Don	Ericaceae
126.	*Geranium nepalense* Sweet	Geraniaceae
127.	*Gentiana stipitata* Edgew	Gentianaceae
128.	*Gleichenia linearis* Bedd	Gleicheniaceae
129.	*Gomphostemma parviflora* Wall	Lamiaceae
130.	*Gonatanthus pumilus* (Don) Engler f. Krause	Araceae
131	*Gouania tiliaefolia* Lamk	Rahmanaceae
132.	*Hedychium acuminatum* Roscoe	Zingiberaceae
133.	*Hedyotis scandens* Roxb.	Rubiaceae
134.	*Hedyotis unicenella* HK & Arn	Rubiaceae
135.	*Hedyotis verticillata* (L) Lamk	Rubiaceae
136.	*Hemidesmus indicus* R.Br. Schult	Asclepiadaceae
137.	*Hemiphriagma heterophyllum* Wall	Scrophulariaceae
138.	*Hibiscus rosa sienensis* L	Malvaceae
139.	*Holarrhena antidysentrica* Roth D.C	Apocynaceae
140.	*Halboellia latifolia* Wall	Lardizabalaceae
141.	*Holmskioldia sanguinea* Retz.	Verbenacea
142.	*Himalomena aromatica* (Roxb.) Schotl.	Araceae
143.	*Houttuynia cordata* Thunb	Saururaceae
144.	*Hoya globulosa* HK.f.	Asclepiadaceae
145.	*Hydrocotyl javanica* Thunb.	Apiaceae
146.	*Hydrocotyl sibthorpioides* Lamk	Apiaceae
147.	*Hypericum japonicum* Thunb	Hypericaceae
148.	*Hypericum monanthemum* HK. f & Th ex Dyer	Hypericaceae
149.	*Hypericum himalaicum* Robson	Hypericaceae
150.	*Hyptanthera stricta* Wt. Arn	Rubiaceae
151.	*Ichnocarpus frutescens* (L) R.Br.	Apocynaceae

152. *Impatiens angustifolia* HK.f	Balsaminaceae
153. *Imperata cylindrica* Beauv	Poaceae
154. *Ipomea quamoclit* (L.)	Convolvulacea
155. *Jasminium amplexicaule* Buch. Ham	Oleaceae
156. *Justicia adhatoda* L.	Acanthaceae
157. *Justicia gendarussa* Burm	Acanthaceae
158. *Kaempferia galanga* L.	Zingiberaceae
159. *Kaempferia rotunda* L.	Zingiberaceae
160. *Kalanchoe pinnata* (Lamk) Pers	Crassulaceae
161. *Kalanchoe laciniata* (L.) Pers	Crassulaceae
162. *Lasia spinosa* (L.) Thw.	Araceae
163. *Leea alata* Edgew	Vitaceae
164. *Leucas plukerneti* (Roth) Spr	Lamiaceae
165. *Leucas indica* L. R.Br ex Vatke	Lamiaceae
166. *Lindernia anagallis* (Burm. f) Pernnell	Scrophulariaceae
167. *Lindenbergia indica* (L.) Kuntze	Scrophulariaceae
168. *Litsea monopetala* (Roxb.) Pers	Lauraceae
169. *Maesa chisia* D.Don	Myrsinaceae
170. *Mahonia acanthifolia* G. Don	Berberidaceae
171. *Mangifera indica* L.	Anacardiaceae
172. *Manihot esculenta* Crantz	Euphorbiaceae
173. *Melastoma malabathricum* L	Melastomtaceae
174. *Mentha arvensis* Linn	Lamiaceae
175. *Meriandra bengalensis* Benth	Lamiaceae
176. *Mesua ferrea* L	Clausiaceae
177. *Microglossa albescens*	Asteraceae

178.	*Micromelum minutum* (Forst. f) Wt & Arn.	Rutaceae
179.	*Mikania micrantha* H.f. K	Asteraceae
180.	*Miliusa roxburghiana* (Griff.) HK.f & Th.	Annonaceae
181.	*Mimosa pudica* L	Mimosaceae
182.	*Mucuna species* Adams	Fabaceae
183.	*Murraya koenigii* Spreng	Rutaceae
184.	*Mussaenda belilla* Buch Ham	Rubiaceae
185.	*Mussaenda roxburghii* HK.f	Rubiaceae
186.	*Myrica esculenta* Buch Ham	Myricaceae
187.	*Myrioneuron nutans* Wall	Rubiaceae
188.	*Naravelia zeylanica* (L.) DC	Ranunculaceae
189.	*Nepenthes khasiana* HK.f	Nepenthaceae
190.	*Nephrolepis cordifolia* (L) Presl	Polypodiaceae
191.	*Nicotiana tabacum* Linn	Solanaceae
192.	*Ocimum basilicum* Linn	Lamiaceae
193.	*Olax acuminata* Wall ex Benth	Olacaceae
194.	*Ophiopogon intermedius* D.Don	Liliaceae
195.	*Ophiorrhiza ochroleuca* HK.f	Rubiaceae
196.	*Oreocnide integrifolia* (Gaud) Miq.	Urticaceae
197.	*Osbeckia stellata* Ker-Gawal	Melastomataceae
198.	*Oxalis corniculata* L.	Oxalidaceae
199.	*Paederia scandens* (Lour) Merr	Rubiaceae
200.	*Pandanus fascicularis* Lamk syn. *P. odoratissimus P. tectorus*	Pandanaceae
201.	*Paris polyphylla* Sm.	Liliaceae

202.	*Parkia roxburghii* G. Don syn. *P. timorina*	Mimosaceae
203.	*Paveta indica* L	Rubiaceae
204.	*Pedilanthus tithymaloides* Poir	Euphorbiacea
205.	*Pegia nitida* Colebr	Anacardiaceae
206.	*Persea gamblei* (King ex Hook f Kost)	Lauraceae
207.	*Phaseolus lunatus* L	Fabaceae
208.	*Pholidota imbricata* (Roxb) Lindl	Orchidaceae
209.	*Pholidota pallida* Lindl	Orichidaceae
210.	*Picrorhiza kurroa* Hook. f	Scrophularia
211.	*Peliosanthes teta* Andr.	Liliaceae
212.	*Piper betle* L	Piperaceae
213.	*Piper mullesus Buch* Ham	Piperaceae
214.	*Plantago major* L.	Plantaginaceae
215.	*Plumbago zeylanica* L	Plumbiginaceae
216.	*Podophyllum hexandrum* Royle Syn. *P. emodi* Wall	Berberidaceae
217.	*Pogostemon purpurascens* Dalz	Lamiacceae
218.	*Polygala persicaraefolia*	Polygalaceae
219.	*Polygonum capitatum* Buch & Ham	Polygonaceae
220.	*Polygonum hydropiper* L.	Polygonaceae
221	*Polygonum strigosum* R.Br.	Polygonaceae
222.	*Portulaca oleraceae* L.	Portulacaceae
223.	*Potentilla fulgens* Wall. Hook	Rosaceae
224.	*Pothos scandens* L	Araceae
225.	*Pouzolzia frondosa* (Don.) Kutze Syn. *P. viminea wedd*	Urticaceae
226.	*Pouzolzia sanguienea* (Bl) Merr	Urticaceae
227.	*Pratia begonifolia* Lindl	Lobeliaceae

228.	*Prunus cerasoides* D. Don	Rosaceae
229.	*Psidum guajava* L.	Myrtaceae
230.	*Punica granatum* L	Punicaceae
231.	*Quercus acutissima* Carrathers	Fagaceae
232.	*Ranunculus diffuses* DC	Ranunculaceae
233.	*Rheum emodi* Wall ex Meissn	Polygonaceae
234.	*Rhus semialata* Merr. Syn. *R. chinensis* Mill.	Anacardiaceae
235.	*Rhynchoglossum obliquumum* Bl.	Gesneriaceae
236.	*Ricinus communis* L.	Euphorbiaceae
237.	*Rosa macrophylla* Lindl.	Rosaceae
238.	*Roscoea purpurea* Smith	Zingiberaceae
239.	*Rubia cordifolia* L.	Rubiacea
240.	*Rubus ellipticus* Sm	Rosaceae
241.	*Rubus niveus* Thunb.	Rosaceae
242.	*Rumex martimus* L.	Polygonaceae
243.	*Sabia lancefolia* Coelebr	Sabiaceae
244.	*Sagina procumbens* L.	Caryophyllace
245.	*Saprosma ternatum* Kurz.	Rubiaceae
246.	*Sarcandra glabra* (Thunb) Nakai	Chloranthaceae
247.	*Satyrium nepalanse* D.Don.	Orchidiaceae
248.	*Schima wallichi* (DC)	Theaceae
249.	*Scoparia dulcis* L.	Scorophulariaceae
250.	*Scutellaria discolor* Colebr.	Lamiaceae
251.	*Securinega virosa* (Roxb. ex Willd)	Euphorbiaceae
252.	*Sedum multicaule* Wall ex Lindl.	Crassulaceae
253.	*Siegesbeckia orientalis* L.	Asteraceae

254.	*Smilacina fusca* Hk. f & Thomas	Liliaceae
255.	*Smilax ovalifolia* Roxb. Syn. *S. macrophylla* Roxb.	Liliacea
256.	*Solanum ferox* L.	Solanaceae
257.	*Solanum macanthum* Dunal Syn. *S. Khasianum*	Solanaceae
258.	*Solanum torvum* Sw.	Solanaceae
259.	*Sonchus asper* (L) Hill	Asteraceae
260.	*Sonerila maculata* Roxb.	Melastomataceae
261.	*Spilanthus paniculata* Wall ex DC.	Asteraceae
262.	*Stachtarpheta jamaiacensis* Vahl	Verbenaceae
263.	*Stauranthera umbrosa* Clarke	Gesneriaceae
264.	*Stephania glabra* Miers	Menispermaceae
265.	*Stephania japonica* Miers Syn. *S. hernandifolia* Walp	Menispermaceae
266.	*Stereospermum chelonoides* DC. Syn. *S. suvaleons* DC.	Bignoniaceae.
267.	*Swertia chirayata* Karat	Gentianaceae
268.	*Swertia dilatata*Wall.	Gentianaceae
269.	*Swertia nervosa* Wall.	Gentianaceae
270.	*Swertia purpurascens* Wall.	Gentianaceae
271.	*Tabernaemontana divericata* (L.) Alston	Apocynaceae
272.	*Taraxacum officinale* Wiggers	Asteracea
273.	*Thalictrum javinicum* Bl.	Ranunculaceae
274.	*Thalictrum minus* L.	Ranunculaceae
275.	*Tinospora cordifolia* (Willd.) Miers	Menispermacea
276.	*Tricosanthes tricuspidata*	Cucurbitaceae
277.	*Urena lobata* L.	Malvaceae

278.	*Valeriana hardwickii* Wall.	Valerianaceae
279.	*Viburnum colebrookianum* Wall	Caprifoliaceae
280.	*Viola biflora* L.	Violaceae
281	*Vitex negundo* L.	Verbenaceae
282.	*Vitex peduncularis* Wall	Verbenaceae
283.	*Vitex trifolia* L.	Verbenaceae
284.	*Vitis barbata* Wall	Vitaceae
285.	*Xeromphis spinosa* (Thunb) Keay	Rubiaceae
286.	*Xylosma longifolium* Clos	Flacourtiaceae
287.	*Zantedeschia aethiopica* (L) Spreng	Araceae
288.	*Zanthoxylum armatum* DC Syn. *Z. alatum* Roxb.	Rutaceae
289.	*Zingiber officinale* Rosc.	Zingiberaceae

DYE YIELDING PLANTS

1.	*Aporusa dioica* Muell-Arg	Euphorbiacea
2.	*Baccaurea ramiflora* Muell-Arg	Euphorbiaceae
3.	*Bischofia javanica* Bl	Euphorbiaceae
4.	*Bixa orellana* Linn	Bixaceae
5.	*Butea minor* Buch-Ham	Fabaceae
6.	*Carthumus tinctorius* Linn	Asteraceae
7.	*Curcuma caesia* Roxb.	Zingiberaceae
8.	*Flemingia stricta* Roxb	Fabaceae
9.	*Garuga pinnata* Roxb.	Buseraceae
10.	*Indigofera dosua* D.Don-Buch Ham	Fabaceae
11.	*Indigofera tinctoria* L	Fabaceae
12.	*Kydia calycina* Roxb.	Malvaceae
13.	*Lithocarpus fenestratus* Roxb.	Fagaceae
14.	*Macrosolen cochinensis* Lout Tiegh	Loranthaceae
15.	*Mallotus philippensis* Muell- Arg.	Euphorbiaceae
16.	*Morinda angustifolia* Roxb	Rubiaceae
17.	*Mammea suriuga* (Buch-Ham ex Roxb.) Kosterm	Clusiacea
18.	*Mussaenda roxburghii* Hook.f	Rubiacea
19.	*Nyctanthes arbortrists* Linn	Oleacea

20.	*Oroxylum indicum* L.Vent	Bignoniaceae
21.	*Onosma emodi* Wall	Boraginaceae
22.	*Pithecellobium angulatum* Benth	Mimosaceae
23.	*Rubia cordifolia* L	Rubiaceae
24.	*Smilax ocreata* A. DC	Liliaceae
25.	*Strobilanthus flaccidifolius* Nees	Acanthaceae
26.	*Symplocos paniculata* (Thunb.) Miq.	Symplocacaceae
27.	*Wood fordia fruticosa* Kurz	Lythraceae

FIBRE YIELDING PLANTS

1.	*Abelmoschus moschatus* Medic	Malvaceae
2.	*Abutilon indicum* L. Sweet	Malvaceae
3.	*Abroma augusta* L.	Sterculiaceae
4.	*Anaphalis busua* (Buch-Ham ex D.Don)	Asteraceae
5.	*Artocarpus chama* Buch-Ham.	Moraceae
6.	*Artocarpus lacucha* Buch-Ham.	Moraceae
7.	*Bauhinia vahlii* Wt & Arn.	Caesalpiniaceae
8 .	*Boehmeria macrophylla* D.Don	Urticaceae
9.	*Boehmeria malabarica* Wedd.	Urticaceae
10.	*Boehmeria nivea* Gaud.	Urticaceae
11.	*Boehmeria platyphylla* D.Don	Urticaceae
12.	*Cissampelos pareira* Linn.	Menispermiaceae
13.	*Chimonobambusa callosa*	Poaceae
14.	*Chimonobambusa intermedia* Nakai	Poaceae
15.	*Corchorus capsularis* Linn.	Tilliaceae
16.	*Corchorus olitorius* Linn.	Tilliaceae
17.	*Daphne papyracea* Wall ex Steud	Thymelaeaceae
18.	*Dendrocnide sinuata* (Blume) Chew.	Urticaceae
19.	*Debregeasia longifolia* (Burn. f) Wedd	Urticaceae
20.	*Dillenia pentagyana* Roxb.	Dilleniaceae
21.	*Edgeworthia gardneri* (Wall.) Meissn	Thymelaeaceae
22.	*Edgeworthia tomentosa* Nakai	Thymelaeaceae

23.	*Flemingia stricta* Roxb.	Fabaceae
24.	*Ficus rumphi* Blume	Moraceae
25.	*Girardinia palmata* (Forsk.) Gaud	Urticaceae
26.	*Grewia disperma* Drummond	Tilliaceae
27.	*Grewia hirsuta* Vahl.	Tilliaceae
28.	*Grewia sapida* Roxb.	Tilliaceae
29.	*Grewia sclerophylla* Roxb. ex G. Don	Tilliaceae
30.	*Hibiscus macrophyllus* Roxb.	Malvaceae
31.	*Hibiscus sabdariffa* L.	Malvaceae
32.	*Indigofera dosua* Buch-Ham	Fabaceae
33.	*Indigofera tinctoria* Linn.	Fabaceae
34.	*Kydia calycina* Roxb.	Malvaceae
35.	*Marsdenia tenacissima* Wight & Arn.	Asclipiadaceae
36.	*Millettia cinerea* Benth	Fabaceae
37.	*Maoutia puya* Wedd	Urticaceae
38.	*Naravellia zeylanica* DC	Ranunculaceae
39.	*Oreocnide frutescens* (Thunb) Miq.	Urticaceae
40.	*Oreocnide integrifolia* Miq.	Urticaceae
41.	*Pericampylus glaucus* (Lamk) Merr.	Menispermaceae
42.	*Pterygota alata* Roxb. Br.	Sterculiaceae
43.	*Polyalthia simiarum* Hook. f & Thomas	Annonaceae
44.	*Sida cordifolia* L.	Malvaceae
45.	*Stephania glabra* (Roxb.) Miers.	Menispermaceae
46.	*Sterculia villosa* Roxb.	Sterculiaceae
47.	*Urtica parviflora* Roxb.	Urticaceae

ESSENTIAL OIL PRODUCING PLANTS

1.	*Acorus calamus* L.	Araceae
2.	*Alstonia scholaris* R.Br.	Apocynaceae
3.	*Aphanamixis polystachya* (Wall) Parker	Meliaceae
4.	*Aquilaria agallocha* Roxb.	Thymeleaceae
5.	*Butea minor* Buch-Ham	Fabaceae
6.	*Cinnamomum camphora* Nees	Lauraceae
7.	*Cinnamomum cecidodaphane* Meissn	Lauraceae
8.	*Curcuma aromatica* Salisb.	Zingiberaceae
9.	*Gaultheria fragrantissima* Wall	Ericaceae
10.	*Hedychium spicatum* Buch-Ham	Zingiberaceae
11.	*Impatiens amplexicaulie* Edgew	Balsaminaceae
12.	*Juglens regia* Linn	Juglandaceae
13.	*Juniperus macropoda* Boiss	Pinaceae
14.	*Lindera neesiana* (Nees) Kurz Benth	Lauraceae
15.	*Litsea cubeba* Pers.	Lauraceae
16.	*Madhuca longifolia* (Koenig.) Macbr	Sapotaceae
17.	*Mesua ferrea* L	Guttiferae/ Clausiaceae
18.	*Nardostachys jatamansi* (Don) DC	Valerianaceae
19.	*Perilla frutescens* (Linn)	Lamiaceae
20.	*Randia dumetorum* Poir.	Rubiaceae
21.	*Ricinus communis* Linn	Euphorbiaceae

SPICES AND CONDIMENT PLANTS

1.	*Allium hookeri* Thw	Liliaceae
2.	*Allium tuberosum* Roth ex Spreng/Roxb	Liliaceae
3.	*Alpinia galanga* (L) Willd	Zingiberaceae
4.	*Cinnamomum bejolghota* S.W.	Lauraceae
5.	*Cinnamomum tamala* S.W.	Lauraceae
6.	*Curcuma longa* L.	Zingiberaceae
7.	*Curcuma amada* L	Zingiberacea
8.	*Eryngium foetidum* L	Apiaceae
9.	*Hedychium coronarium* Koeing ex Retz	Zingiberaceae
10.	*Hedychium spicatum* Buch-Ham ex Smith	Zingiberaceae
11.	*Illicium griffthii* HK.f & Th.	Magnoliaceae
12.	*Murraya koenigii* (L) Spreng	Rutaceae
13.	*Zanthoxylum armatum* DC	Rutaceae

GUM YIELDING PLANTS

1.	*Acacia farnesiana* Willd.	Mimosaceae
2.	*Adansonia digitata* Linn.	Bombacaceae
3.	*Albizia lebbeck* Benth.	Mimosaceae
4.	*Azadirachta indica* A. Juss.	Meliaceae
5.	*Butea monosperma* (Lamk.) Taub.	Fabaceae
6.	*Cymopsis tetragonolobus* L.	Fabaceae
7.	*Delonix regia* Rafin.	Caesalpiniaceae
8.	*Gardenia turgida* Roxb.	Rubiaceae
9.	*Grevillea robusta* A. Cunn.	Proteaceae
10.	*Lagarstroemia parviflora* Roxb.	Lythraceae
11.	*Mitragyna parviflora* Roxb. (Korth)	Rubiaceae
12.	*Streculia urens* Roxb.	Sterculiaceae
13.	*Terminalia bellirica* Roxb.	Combretaceae

ETHONOTOXIC PLANTS

1.	*Albizia procera* (Roxb) Benth	Mimosaceae
2.	*Acacia pennata* (L) Willd	Mimosaceae
3.	*Anamirta paniculata* Colebr	Menispermaceae
4.	*Barringtonia acutangula* (L) Gaertn	Barringtoniaceae
5.	*Buddleja asiatica* Lour	Buddlejaceae
6.	*Canthium dicoccum* Merrill	Rubiaceae
7.	*Cassia memosoides* L.	Caesalpiniaceae
8.	*Croton caudatus* Geiseler	Euphorbiaceae
9.	*Croton tiglium* Linn	Euphorbiaceae
10.	*Derris elliptica* Benth	Fabaceae
11.	*Derris scandens* Benth	Fabaceae
12.	*Engelhardtia polystachya* Radik	Juglandaceae
13.	*Eupatorium odoratum* L	Asteraceae
14.	*Ficus elastica* Roxb	Moraceae
15.	*Ficus silhetensis* Miq	Moraceae
16.	*Gynocardia odorata* R.Br	Flacourtiaceae
17.	*Hydnocarpus kurzii* Warb	Flacourtiaceae
18.	*Maesa indica* Wall	Myrsinaceae
19.	*Milletia auriculata* Baker ex Brandis	Fabaceae
20.	*Milletia pachycarpa* Benth	Fabaceae

21.	*Randia dumetorum* Poir	Rubiaceae
22.	*Randia fasciculate* DC	Rubiaceae
23.	*Schima wallichii* DC Korth	Theaceae
24.	*Tepherosia candida* DC	Fabaceae
25.	*Trevesia palmate* (Roxb.) Vis	Araliaceae
26.	*Uraria crinita* Desv	Fabaceae
27.	*Vibrnum corlifolum* HK.f & Th	Caprifoliaceae
28.	*Viburnum foetidum* Wall	Caprifoliaceae
29.	*Zanthoxylum acanthopodium* DC.	Rutaceae

RESIN YIELDING PLANTS

1.	*Altingia excelsa* Noronha	Altingiaceae
2.	*Baliospermum montanum* Muell. Arg	Euphorbiacea
3.	*Calotropis gigentia* (L.) R.Br.	Asclepidaceae
4.	*Cannabis sativa* Linn	Cannabinaceae
5.	*Garcina mangostana* Linn	Guttiferae/ Clusiaceae
6.	*Guaiacum officinale* Linn	Zygophyllaceae
7.	*Ipomoea eriocarpa* R.Br.	Convolvulaceae
8.	*Justicia adhatoda* L.	Acanthaceae
9.	*Leonotis nepatefolia* R. Br.	Lamiaceae
10.	*Mallotus philippensis* Muell- Arg	Euphorbiaceae
11.	*Mirabilis jalapa* Linn	Nyctanginaceae
12.	*Pinus merkusii* Jungh & Devriesa	Pinaceae
13.	*Plumeria rubra* Linn	Apocynaceae
14.	*Podophylum hexandrum* Royle	Berberidaceae

TANIN YIELDING PLANTS

1.	*Acacia farnesiana* L. Willd	Mimosaceae
2.	*Acacia torta* Linn	Mimosaceae
3.	*Albizia lebbek* Benth	Mimosaceae
4.	*Alnus nepalensis* D.Don	Betulaceae
5.	*Bischofia javanica* Bl	Euphorbiaceae
6.	*Careya arborea* Roxb.	Lecythidaceae
7.	*Cassia obtusifolia* Linn	Caesalpiniaceae
8.	*Casurina equisetifolia* Linn	Casuarinaceae
9.	*Crataeva adansonii* Linn	Cappridaceae
10.	*Diospyros peregrina* (Gaertn) Gurke	Ebenaceae
11.	*Dodonaea viscosa* Linn	Sapindaceae
12.	*Emblica officinalis* Garten	Euphorbiaceae
13.	*Garcinia mangostana* Linn	Guttiferae/ Clusiaceae
14.	*Grevillea robusta* A. cunn	Proteaceae
15.	*Hymenodictyon excelsum* Wall	Rubiaceae
16.	*Jatropha curcas* Linn	Euphorbiaceae
17.	*Lagerstroemia parviflora* Roxb.	Lythraceae
18.	*Lannea coromondelica* (Houtt) Merill	Anacardiac eae
19.	*Mallotus phillippensis* Muell-Arg	Euphorbiaceae

20.	*Michelia doltsopa* Buch –Ham	Magnoliaceae
21.	*Nerium indicum* Mill	Apocynaceae
22.	*Nyctanthes arbortristis* Linn	Oleaceae
23.	*Pithecellobium dulce* Benth	Mimosaceae
24.	*Polygonum plebeium* R.Br.	Polygonaceae
25.	*Spondias pinnata* (L.f) Kurz.	Anacardiaceae
26.	*Syzygium cuminii* Skeel.	Myrtaceae
27.	*Tamarindus indica* Linn	Caesalpiniaceae
28.	*Terminalia arjuna* (Roxb) Wt. Arn	Combretaceae
29.	*Woodfordia fruiticosa* Kurz	Lythraceae

TIMBER PLANTS

1.	*Acacia catechu* (L.f) Willd	Mimosaceae
2.	*Acacia feruginea* (L) Willd	Mimosaceae
3.	*Acacia leucophloea* Willd	Mimosaceae
4.	*Acacia sundra* DC	Mimosaceae
5.	*Adenia sessifolia*	Rubiaceae
6.	*Adina cordifolia* (Roxb) HK.f	Rubiacea
7.	*Aegle marmelos* (L) Corr	Rutaceae
8.	*Albiazia lebbek* (L) Benth	Mimosaceae
9.	*Allaenthus excelsa* Thew.	Moraceae
10.	*Alstonia scholaris* (L) R. Br.	Apocynaceae
11.	*Amoora wallichi* King Wight & Arn	Meliaceae
12.	*Amoora rohituka* Weight and & Arn	Meliaceae
13.	*Anogeissus acuminata* Wall ex Bedd	Combretaceae
14.	*Anogeissus latifolia* Wall ex Bedd	Combretaceae
15.	*Anthocephalus cadamba* (Roxb) Miq	Rubiaceae
16.	*Aquillaria agallocha* Roxb.	Thymelaceae
17.	*Azaddiracta indica* A. Juss.	Meliaceae
18.	*Barringtonia acutangla* (L) Gaertn	Lacythidaceae
19.	*Bauhinia malabarica* (Roxb) Benth	Caesalpiniaceae
20.	*Bauhinia purpurea* Roxb.	Caesalpiniaceae

21.	*Bauhinia racemosa* Lamk	Caesalpiniaceae
22.	*Bauhinia retusa* Roxb.	Caesalpiniaceae
23.	*Bischofia javanica* Bl	Euphorbiaceae
24.	*Boswellia serrata* Roxb. Ex Coelbr	Burseraceae
25.	*Buchanania lanzan* Spereng	Anacardiaceae
26.	*Butea monosperma* Kuntz	Fabaceae
27.	*Calicarpa arborea* Roxb	Verbenaceae
28.	*Camellia sinensis* (Linn) O. Kuntz	Theaceae
29.	*Canarium bengalense* Roxb.	Burseraceae
30.	*Casearia gravelons* Dalz	Samydaceae
31.	*Casearia tomentosa* Roxb.	Samydaceae/ Flacourtiaceae
32.	*Cassia fistula* L	Caeslpiniaceae
33.	*Cassia simea* Lam	Caesalpiniaceae
34.	*Chloroxylon swietenia* DC	Rutaceae
35.	*Cinnamomum cecidodaphne* Meissn	Lauraceae
36.	*Cleistanthus collinus* Roxb. Benth & Hook	Euphorbiaceae
37.	*Cochlospermum tomentosum* Kunth	Chochliospermaceae
38.	*Cordia myxa* Roxb.	Boraginaceae
39.	*Cynometra polyandra* Roxb	Caesalpiniaceae
40.	*Dalbergia latifolia* Roxb	Fabaceae
41.	*Dichrostachys cinerea* Wight & Arn	Araceae
42.	*Dillenia pentagyna* Roxb.	Dilleniaceae
43.	*Diospyros chloroxylon* Roxb.	Ebenaceae
44.	*Diospyros embryopteris* Pers	Ebenaceae
45.	*Diospyros melanoxylon* Roxb.	Ebenaceae

46.	*Dipterocarpus macrocarpus* Vesque	Dipterocarpaceae
47.	*Dipterocarpus turbinatus* Gaertn.f	Dipterocarpaceae
48.	*Dolichandrone rheedii* Seem	Bignoniaceae
49.	*Elaeodendron glaucum* Pers	Celastraceae
50.	*Emblica officinalis* Gaertn	Euphorbiaceae
51.	*Erythrina indica* Lamk	Fabaceae
52.	*Erythrina suberosa* Roxb.	Fabaceae
53,	*Eucalyptus citriodra* Hk	Myrtaceae
54.	*Eugenia jumbolana* Wall Syn. *Syzygium jambolana,* *Syzygium cuminii* Skeels	Myrtaceae
55.	*Feronia limonia Swingle* Syn. *Feronia elephantum* Correa	Rutaceae
56.	*Ficus religiosa* L	Moraceae
57.	*Ficus rumphii* Blume	Moraceae
58.	*Flacourtia ramonotchi* L. Herit	Flacourtaceae
59.	*Gmelina arborea* Roxb.	Verbenaceae
60.	*Gyrocarpus americanus* Jacq	Hernandiaceae
61.	*Gyrocarpus jacquini* Garten	Hernandiaceae
62.	*Hardwickia pinnata* Roxb.	Caesalpinniaceae
63.	*Holarrhena antidysentrica* Wall	Apocynaceae
64.	*Holoptelea integrifolia* Roxb. Planch	Ulmaceae
65.	*Hymenodictyon excelsum* Roxb. Wall	Rubiaceae
66.	*Ixora arborea* Roxb. ex Sm	Rubiaceae
67.	*Kydia calcycina* Roxb.	Malvaceae
68.	*Lannea grandis* Engl	Anacardiaceae
69.	*Lagerstroemia parviflora* Roxb.	Lythraceae
70.	*Litsea polyantha* Juss	Lauraceae

71.	*Machilus gamblei* (King ex HK.f)	Lauraceae
72.	*Macranga peltata* (Roxb) Muell-Arg.	Euphorbiaceae
73.	*Madhuca indica* J.F. Gambel	Sapotaceae
74.	*Mallotus philippensis* (Lamk) Muell-Arg	Euphorbiaceae
75.	*Mangifera sylvatica* Roxb.	Anacardiaceae
76.	*Manihot esculenta* Crantz	Euporbiaceae
77.	*Michelia champaca* L.	Magnoliaceae
78.	*Millingtonia hortensis* Linn.f	Bignoniaceae
79.	*Mimusops elengi* L	Sapotaceae
80.	*Mitragyna parvifolia* (Roxb) Korth	Rubiaceae
81.	*Morinda tinctoria* Roxb. Syn. *M. corcia* Buch- Ham	Rubiaceae
82.	*Neohouzeaua dullooa* (Gamble)	Poaceae
83.	*Nyctanthes arbortristis* L	Olaceae
85.	*Odina wodier* Roxb.	Anacardiaceae
86.	*Oroxylem indicum* (L) vent	Bignoniaceae
87.	*Oxytenanthera nigrocilliata* Munro	Poaceae
88.	*Phoebe goalparensis* Nees	Lauraceae
89.	*Phonix sylvestris* Roxb	Palmaceae
90.	*Plectronia dydima* L	Rubiaceae
91.	*Polyalthia cerasoides* (Roxb.) Bedd	Anoniaceae
92.	*Polyalthia longifolia* Thew	Annoniaceae
93.	*Pongamia pinnata* (L.) Pierre	Fabaceae
94.	*Premna bengalensis* C.B. Clarke	Verbenaceae
95.	*Premna tomentosa* Willd.	Verbenaceae
96.	*Prosopis cineraria* (L.) Druce Syn. *P. specigera* L	Mimosaceae

97	*Pterocarpus indicus* Willd.	Fabaceae
98.	*Pterocarpus marsupium* Roxb.	Fabaceae
99.	*Pterygota alata* (Roxb.) R.Br.	Sterculiaceae
100.	*Randia dumetorum* Poir.	Rubiaceae
101.	*Rhododendron arboreum* Sm.	Ericaceae
102.	*Saccopetalum tomentosum* Hook. f. & Thomas	Anonaceae
103.	*Salmalia malabarica* DC. Schott *Bomba malabarica, B, ceiba*	Bombacaceae
104.	*Santalum album* L	Santalaceae
105.	*Sapindus emarginatus* Vahl	Sapandiacea
106.	*Schima wallichi* (DC.) Korth	Theaceae
107.	*Schleichera oleosa* (Lour.) Oken	Sapindaceae
108.	*Schrebera swietenioides* Roxb.	Oleaceae
109.	*Semecarpus anacardium* L.f	Anacardiaceae
110.	*Shorea robusta* Gaertn.f	Dipterocarpaceae
111.	*Soyamida febrifuga* (Roxb.) A. Juss	Meliaceae
112.	*Spondias mangifera* Willd	Anacardiaceae
113.	*Sterculia Urens* Roxb.	Sterculiaceae
114.	*Sterculia villosa* Roxb.	Sterculiaceae
115.	*Stereospermum chelonoides* (L.f.) DC	Bignoniaceae
116.	*Stereospermum personatum* Hassk	Bignoniaceae
117.	*Stereospermum suaveolens* (Roxb) DC	Bignoniaceae
118.	*Swietenia mahogoni* Jacq	Meliaceae
119.	*Strychonos potatorum* L.	Loganiaceae
120.	*Tamarindus indica* L	Caesalpiniacea
121.	*Termanilia arjuna* (Roxb) Wt ex Arn	Combretaceae
122.	*Terminalia bellirica* (Gaertn) Roxb.	Combreteceae

123.	*Terminalia chebula* Retz	Combreteaceae
124.	*Termnalia tomentosa* WT & Arn	Combreteaceae
125.	*Tetrameles nudiflora* R.Br.	Tetramelaceae
126.	*Trema orientalis* Blume	Ulmaceae
127.	*Trewia nudiflora* L	Euphorbiaceae
128.	*Vitex peduncularis* Wall ex Schauer	Verbenaceae
129.	*Wrightia arborea* Roxb. Roem Scultz	Apocynaceae
130.	*Xylia dolabriformis* Benth	Mimosaceae
131.	*Xylia xylocarpa* (Roxb) Taub	Mimosaceae
132.	*Zanthoxylem budrunga* (Roxb) DC Syn. *Z. armatum* DC	Rutaceae

FOOD AND EDIBLE PLANTS

1.	*Acetephila excelsa* (Dalz) Muell Arg	Euphorbiaceae
2.	*Actinidia chinensis* Planch	Actinidiacea
3.	*Aegle marmelos* (L) Corr	Rutaccac
4.	*Aesandra butyraceae* (Roxb)	Sapotaceae
5.	*Aglaia edulis* A. Gray	Meliaceae
6.	*Alpinia galanga* Willd	Zingiberaceae
7.	*Alpinia malaccensis* Rosc	Zingiberaceae
8.	*Alpinia nigra* (Gaertn)	Zingiberacea
9.	*Alpinia rubra* (Roxb) Radik	Zingiberacea
10.	*Alpinia speciosa* K. Schum	Zingiberacea
11.	*Amaranthus caudatus* Linn	Amaranthacrar
12.	*Amorphophallus campanulatus* (Roxb) Bl.	Araceae
13.	*Ampelocissus barbata* Wall	Vitaceae
14.	*Angiopteris evecta* (Forst) Hoffm	Marattiaceae
15.	*Antidesma acuminatum* Wall	Euphorbiaceae
16.	*Antidesma diandrum* (Roxb) Hayne ex Roth	Euphorbiaceae
17.	*Antidesma thwitesianum* Muell Arg	Euphorbiacea
18.	*Aporusa dioica* Muell-Arg	Euphorbiaceae
19.	*Ardisia crispa* Thunb DC	Myrsinaceae

20.	*Ardisia griffithi* Clarke	Myrsinacea
21.	*Ardisia polycephala* Wall	Myrsinacea
22.	*Ardisia solanaceae* Roxb.	Myrsinaceae
23.	*Artocarpus heterophyllus* Lamk	Moracea
24.	*Asparagus racemosus* Willd	Liliaceae
25.	*Asystassia neesiana* (Wall) Nees	Acanthaceae
26.	*Averroha carambola* Linn	Oxalidacea
27.	*Baccurea sapida* Muell. Arg Syn *B. ramifolia* Lour	Euphorbiaceae
28.	*Bambusa bambos* (L) Voss	Poaceae
29.	*Bambusa tulda* Roxb.	Poaceae
30.	*Bauhinia purpurea* L	Caesalpiniacea
31.	*Benincassa hispida* (Thunb)	Cucurbitaceae
32.	*Bombax ceiba* L	Bombacaceae
33.	*Buddleja asiatica* Lour	Loganaceae
34.	*Bupleurum falcatum* L.	Apiaceae
35.	*Calamus rotang* L.	Areaceae
36.	*Capsicum annum* Linn	Solanaceae
37.	*Casearia esculenta* Roxb.	Samydaceae
38.	*Castanopsis tribuloides* (Smith) A.D.C.	Fagaceae
39.	*Celtis australis* Linn	Ulmaceae
40.	*Chenopodium album* Linn	Chenopodiaceae
41.	*Chlorophytum tuberosum* Baker	Liliaceae
42.	*Cirsium lipskyi* Petrak	Asteraceae
43.	*Citrus hystrix* DC	Rutaceae
44.	*Citrus medica* Linn	Rutaceae
45.	*Clausena dentata* (Willd) Roem & Schult	Rutaceae

46.	*Clerodendrum colebrookianum* Walp	Verbenaceae
47.	*Clerodendrum serratum* L. Moon	Verbenaceae
48.	*Codonopsis ovata* Benth	Campanulaceae
49.	*Colocasia esculenta* (Linn) Schott	Araceae
50.	*Conocephalus suaveolens* Bl	Moraceae
51.	*Coptis teta* Wall	Ranunculaceae
52.	*Cordia wallichii* G. Don	Boraginaceae
53.	*Cordia nepalensis* Hook & Thomas	Boraginaceae
54.	*Cornus capitata* Wall ex Royle	Cornaceae
55.	*Costus speciosus* (Koening) Sm	Zingiberaceae
56.	*Crataegus oxycantha* Linn	Rosaceae
57.	*Cucumis sativus* Linn	Cucurbitaceae
58.	*Cucurbita maxima* Duch	Cucurbitaceae
59.	*Curcuma zedoria* Rosc.	Zingiberaceae
60.	*Cyanotis tuberosa* Schult.f	Commelinaceae
61.	*Cyathocalyx martabanicus* Hk. F. Th	Annonaceae
62.	*Dendrocalmus hamiltoni* Nees & Arn	Poaceae
63.	*Dichopsis polyantha* Benth & HK.f Syn. *Palaquium polyanthus* (Wall Baill)	Sapotaceae
64.	*Dillenia indica* Linn	Dillieniaceae
65.	*Dillenia pentagyna* Roxb.	Dillieniaceae
66.	*Diospyros kaki* Linn. f	Ebenaceae
67.	*Dioscorea bulbifera* Linn	Dioscoriaceae
68.	*Dioscorea glabra* Roxb.	Dioscoriaceae
69.	*Dioscorea pentaphylla* Linn	Dioscoriaceae
70.	*Dioscorea puber* Blume	Dioscoriaceae
71.	*Dioscorea trinervis* Roxb. ex Prain & Bark	Dioscoriaceae

72.	*Dioscorea wallichi* HK.f	Dioscoriaceae
73.	*Docynia hookeriana* Decne	Rosaceae
74.	*Donella roxburghii* (G.Don)	Sapotaceae
75.	*Duchesenea indica* (Andr) Focke	Rosaceae
76.	*Ehretia acuminata* R.Br.	Boraginaceae
77.	*Elaeocarpus varunna* Buch-Ham	Elaeocarpaceae
78.	*Elaeagnus latifolia* Linn	Elaeagnaceae
79.	*Elatostema dissectum* Wedd	Urticaceae
80.	*Elatostema platyphyllum* Wedd	Urticaceae
81.	*Eleocharis dulcis* Trin	Cyperaceae
82.	*Eleusine corcona* Gaertn	Poaceae
83.	*Embelia gamblei* Kurz	Myrsinaceae
84.	*Embelia sessifolia* Kurz	Myrsinaceae
85.	*Emblica officinalis* Gaertn	Euphorbiaceae
86.	*Ensete superbum* (Roxb.) Cheeseman	Musaceae
87.	*Entada phaseoloides* Merill	Fabaceae
88.	*Entada pursetha* DC	Fabaceae
89.	*Eryngium foetidum* L	Apiaceae
90.	*Erioglossum rubiginosum* Blume	Sapindaceae
91.	*Eriosema chinense* Vog	Fabaceae
92.	*Eugenia formosa* Wall	Myrtaceae
93.	*Euginia kurzii* Duthie	Myrtaceae
94.	*Eulophia campestris* Wall	Orchidaceae
95.	*Euryale ferox* Salisb	Nymphaeaceae
96.	*Evodia froxinifolia* (D.Don) Hk.F	Rutaceae
97.	*Fagopyrum cymosum* Meissn	Polygonaceae
98.	*Feronia limonia* Swingle	Rutaceae

99.	*Ficus auriculata* Lour	Moraceae
100.	*Ficus hirta* Vahl	Moraceae
101.	*Ficus hispida* Linn	Moraceae
102.	*Fissistigma Polyanthum* Merrill	Anonaceae
103.	*Flacourtia jangomanus* Raeuschl	Flacourtiaceae
104.	*Flemingia vestita* Benth	Fabaceae
105.	*Galingsopa quadriradiata* Ruiz & Pavon Syn. *G. parviflora* Cav	Asteraceae
106.	*Garcinia atrovirdis* Griff	Clausiaceae
107.	*Garcinia lanceaefolia* Roxb.	Clausiaceae
108.	*Garcinin paniculata* Roxb.	Clausiaceae
109.	*Garcinia xanthochymus* Hk.f	Clausiaceae
110.	*Gardenia campanulata* Roxb.	Rubiacrar
111.	*Gardenia latifolia* Ait	Rubiaceae
112.	*Gaultheria fragrantissima* Wall	Ericaceae
113.	*Gelonium multiflorum*	Euphorbiaceae
114.	*Gironniera cuspidata* Bl. Kurz	Ulmaceae
115.	*Gnetum gnemon* Linn	Gnetaceae
116.	*Haematocarpus thomsonnii* Miers	Menispermaceae
117.	*Hodgsonia hetrociliata* Hk.f & Th	Cucurbitaceae
118.	*Hodgsonia macrocarpa* (Bl) Cogn	Cucurbitaceae
119.	*Horsfieldia amygdalina* (Wall) Warb	Myristicaceae
120..	*Houttuynia cordata* Thunb	Saururaceae
121.	*Hovenia dulcis* Thunb	Rahmanacea
122.	*Hygrophila salcifolia* Nees	Acanthaceae
123.	*Illicium griffithii* Hk.f.& Thomas	Magnoliaceae
124.	*Impatiens amplexicaule* Edgew	Balsaminaceae

125.	*Indigofera dosua* Buch Ham ex D. Don	Fabaceae
126.	*Ipomoea aquatica* Forsk	Convolvulaceae
127.	*Kydia calycina* Roxb.	Malvaceae
128.	*Lasia spinosa* L. Thw	Araceae
129.	*Leea asiatica* (L) Ridsdale Syn *Leea crispa*	Vitaceae
130.	*Leea indica* (Burm. f) Meir	Vitaceae
131.	*Lepisanthes rubiginosa* Roxb.	Sapindaceae
132.	*Lindra neesiana* Benth	Lauraceae
133.	*Litsea cubeba* (Lour) Pers	Lauraceae
134.	*Litsea glutinosa* (Lour) Robins	Lauraceae
135.	*Litsea kingii* Hk.f	Lauraceae
136.	*Lysmachia candida* Lindl	Primulaceae
137.	*Madhuca longifolia* var. *latifolia* (Koen) Macbr	Sapotaceae
138.	*Maesa argentea* Wall	Myrsinaceae
139.	*Maesa indica* Wall	Myrsinaceae
140.	*Maesa montana* DC	Myrsinaceae
141.	*Mahonia acanthifolia* G. Don	Berberidaceae
142.	*Malchara capitata* (L)	Malvaceae
143.	*Malva verticillata* L	Malvaceae
144.	*Mangifera indica* L	Anacardiaceae
145.	*Mangifera sylvatica* Roxb.	Anacardiaceae
146.	*Medinilla rubicunda* Bl	Melastomaceae
147.	*Melastoma malabathricum* L	Melastomaceae
148.	*Meliosma pinnata* Hallier f	Sabiaceae
149.	*Melodinus monogynus* Roxb.	Apocynaceae
150.	*Merremia umbellata* (Linn)	Convolulaceae

151 . *Mesua ferrea* L	Clausiaceae
152. *Meyna spinosa* Roxb.	Rubiaceae
153. *Monochoria hastaefolia* Presl	Pontederiaceae
154. *Morinda umbellata* L	Rubiaceae
155. *Moringa oleifera* Lamk	Moringaceae
156. *Morus alba* L	Moraceae
157. *Mucuna prurita* (L) DC	Fabaceae
158. *Murraya koenigii* (L) Spreng	Rutaceae
159. *Mussaenda frondosa* L	Rubiaceae
160. *Mussaenda glabra* Vahl	Rubiaceae
161. *Mussaenda roxburghii* Hk.f	Rubiaceae
162. *Myrica eaculenta* Buch-Ham ex D.Don	Myricaceae
163. *Myrsine capitellata* Wall	Myrsinaceae
164. *Natsiatum herpeticum* Buch-Ham	Icaciniaceae
165. *Nelumbo nucifera* Gaertn	Nymphaceae
166. *Nymphoides indicum* (L) Kuntz	Nymphaceae
167. *Oenanthe javanica* (Bl) DC	Apiaceae
168. *Olax acuminata* Wall ex Benth	Olacaceae
169. *Olax nana* Wall ex Benth	Olacaceae
170. *Oxalis martiana* Zucc	Oxalidaceae
171. *Parthenocissus himalyana* (Royle) Planch	Vitaceae
172. *Paedaria foetida* L.	Rubiaceae
173. *Parkia roxburghii* G. Don Syn. *P. timorina* (A.D.C.) Merill	Mimosaceae
174. *Pavetta subcapitata* HK.f	Rubiaceae
175. *Pegia nitida* Coelebr	Anacardiaceae
176. *Pennisetum americanum* (L) Leeke	Poaceae

177.	*Persea kurzi* (King ex HK.f) Kosterm	Lauraceae
178.	*Phaseolus lunata* Roxb.	Fabaceae
179.	*Phonenix acaulis* Buch-Ham	Palmaceae/ Arecaceae
180.	*Phytolacca acinosa* Roxb.	Phytolaccaceae
181.	*Picris hieraciodes* L	Asteraceae
182.	*Pinanga hookeriana* Bece	Palmaceae/ Arecaceae
183.	*Polygonatum multiflorum* (L) Ait	Liliaceae
184.	*Polygonatum verticiliatum* (L) Ait	Liliaceae
185.	*Polygonum runicinatum* Buch Ham	Polygonaceae
186.	*Pouzolzia Viminea* Wedd	Urticaceae
187.	*Premna latifolia* Roxb	Verbenaceae
188.	*Protium serratum* Wall ex Colebr	Burseraceae
189.	*Prunus cera soides* D. Don	Rosaceae
190.	*Prunus cornuta* (Royle) Steud	Rosaceae
191.	*Pteridium aquilinium* Kuhn	Polypodaceae
192.	*Pterygota alata* (Roxb) R.Br.	Sterculaceae
193.	*Pyrus pashia* Buch-Ham ex D. Don	Rosaceae
194.	*Pyrularia edulis* (Wall) A.D.C.	Santalaceae
195.	*Randia spinosa* Poir	Rubiaceae
196.	*Randia uliginosa* (Retz) DC	Rubiaceae
197.	*Rhus javanica* L	Anacardiaceae
198.	*Rhynochotechum ellipticum* A.DC	Gesneriaceae
199.	*Ribes glaciale* Wall	Saxifragaceae
200.	*Rorippa indica* (L) Heirn	Cruciferae
251.	*Rosa gigantia* Collett	Rosaceae
202.	*Rourea minor* (Gaertn) Alston	Connaraceae

203. *Rubus biflorus* Buch-Ham ex Sm	Rosaceae
204. *Rubus rogosus* Sm	Rosaceae
205. *Rumex martimus* L.	Polygonaceae
206. *Saggittaria sagittifolia* L Syn. *S. trifolia*	Alismaceae
207. *Salacia roxburghii* Wall	Hippocrateaceae
208. *Sambucus javanica* Reinw ex Bl Syn. *S. hookeri Rehder*	Caprifoliaceae
209. *Sarcochlamys pulcherrima* Gaudich	Urticaceae
210. *Saussurea cerea* Griff ex Dyer	Asteraceae
211. *Smilacina fusca* Wall	Liliaceae
212. *Solanum barbisetum* Nees	Solanaceae
213. *Solanum crassipetalum* Wall	Solanaceae
214. *Solanum kurzii* Brace ex Prain	Solanaceae
215. *Solanum melongena* L	Solanaceae
216. *Solena heterophylla* Lour	Cucurbitaceae
217. *Sorbus aucuparia* L	Rosaceae
218. *Sterculia guttata* Roxb.	Sterculiaceae
219. *Sterculia indica* Merill	Sterculiaceae
220. *Stixis suaveolens* (Roxb) Pierre	Capparidiaceae
221. *Strebulus indicus* (Bur) Corner	Moraceae
222. *Syzygium aqueum* Alston	Myrtaceae
223. *Syzygium cuminii* {Linn.} Skeel	Myrtaceae
224. *Tetrastigma lanceolarium* (Roxb) Planch	Vitaceae
225. *Tricosanthes dioic* Roxb.	Cucurmitaceae
226. *Trigonostemon semperforeus* Blume	Euphorbiaceae
227. *Typha aungustifolia* Watt	Typhaceae
228. *Uraria cordata* (Dunal) Alston	Fabaceae

229.	*Vaccinium donianum* Wight	Vaccinniaceae
230.	*Vaccinium griffthianum* Wight	Vacciniaceae
231.	*Vaccinium serratum* Wight	Vacciniaceae
232.	*Vaccinium sprengelli* Wall	Vacciniaceae
233.	*Viburnum colebrookianum* Wall ex C.B.Clarke	Caprifoliaceae
234.	*Viburnum continifolium* Don	Caprifoliaceae
235.	*Viburnum grandiflorum* Wall	Caprifoliaceae
236.	*Vigna pilosa* Baker	Fabaceae
237.	*Vigna sinensis* (L) Savi ex Hask	Fabaceae
238.	*Vigna umbellata* (Thunb) Ohwi & Ohashi	Fabaceae
239.	*Vigna vexillata* (Benth) A. Rich	Fabaceae
240.	*Vitis lanata* Roxb.	Vitaceae
241.	*Willughebia edulis* Roxb.	Apocynaceae
242.	*Ximenia americana.* L	Olacaceae
243.	*Zanthoxylum oxyphyllum* Edgew	Rutaceae
244.	*Zingiber zerumbet* (L) Rosc ex Smith	Zingiberaceae
245.	*Zizyphus apetla* H.K.f	Rhamanaceae
246.	*Zizyphus maurtiana* Lam.	Rhamanaceae
247.	*Zizyphus rotundifolia* Lam.	Rhamanaceae

LIST OF THE LEGUMINOUS PLANTS WHICH HAVE BEEN LOCATED IN NORTH EAST

	Plant Name	Edible/ Non-Edible	Medicinal Use Yes/No
1.	*Abrus fruticulosus* Wall ex W&A	N.E.	Yes
2.	*Abrus prectorius* L	N.E.	Yes
3.	*Abrus pulchelus* Wall ex Thew	N.E	Yes
4.	*Acacia catechu* (L) Willd	N.E	Yes
5.	*A. Concina* (Willd) D.C.	N.E	Yes
6.	*A. farensiana* (L) Willd	N.E.	Yes
7.	*A. mearnsil* Willd	1	1
8.	*A. melanoxylon* R.Br.	1	1
9.	*A. Moliformis* Griseb	1	1
10.	*A.nilotica* L. Willd	N.E.	Yes
11.	*A. pennata* Willd	N.E.	1
12.	*A. pruinescnsis* Kurz	Yes	1
13.	*Adenthera pavonina* L.	N.E	Yes,
14.	*Aeschynomene aspera* L	N.E.	Yes
15.	*A. indica* L	N.E.	Yes
16.	*Albizzia arunachaelensis* Sahani and Nathani	1	1

	Plant Name	Edible/ Non-Edible	Medicinal Use Yes/No
17.	*A. chinensis* (Osbeck) Merril	N.E.	Yes
18.	*A.julibrussin* Duraz	N.E.	Yes
19.	*A. lebbeck* (L) Benth	N.E.	Yes
20.	*A. lucidior* (Steud) Nielsen	N.E.	Yes
21.	*A. myriophylla* Benth	N.E	Yes
22.	*A. odoratissima* Benth	1	1
23.	*A. smithiana* Prains	1	1
24.	*A. stipulata* Bovian	1	1
25.	*Alysicarpus rogosus* (Willd) DC	E	1
26.	*A. tetragonolobus* Edgew	1	1
27.	*A.vaginalis* (Linn) DC	1	1
28.	*Atylosia goensis* Dalz	N.E.	Yes
29.	*A. scaraboides* (L) Benth	N.E	Yes
30.	*Bauhinia acuminata* L	N.E.	Yes
31.	*B. anguina* Roxb.	1	1
32.	*B. khasiana* Baker	1	1
33.	*B. malabarica* Roxb.	N.E.	1
34.	*B. purpurea* L.	N.E.	Yes
35.	*B. tenuiflora* Wall ex C.B Clarke	N.E.	Yes
36.	*B. variegata* Linn	N.E.	Yes
37.	*B. wallichii* Macbeth	1	1
38.	*Butea buteformis* (Voigt) Gresson and Long	N.E.	Yes
39.	*B. monosperma* (Lamk) Taub	N.E.	Yes
40.	*B. parviflora* Roxb	N.E.	Yes
41.	*B. purpurea* (Benth ex Baker)	N.E.	Yes

	Plant Name	Edible/ Non-Edible	Medicinal Use Yes/No
42.	*Caesalpinia bonduc* (L). Roxb.	N.E.	Yes
43.	*C. decapitata* (Roth) Alst	N.E.	Yes
44.	*C. pulcherimma* (L) Sw	N.E.	Yes
45.	*Callindra haematocephala* Hassk	N.E	Yes
46.	*C. umberosa* (Wall) Benth	1	1
47.	*Canavalia ensiformis* (L) DC	E	Yes
48.	*C. gladiata* (Jacq.) DC	E	Yes
49.	*Cassia fistula* Linn.	N.E.	Yes
50.	*C. floribunda* Cav	1	1
51.	*C. Javanica* L	1	1
52.	*C. occidentalis* L	N.E.	Yes
53,	*C. renigera* Wall ex Benth	1	1
54.	*C. tora* (Linn)	N.E	Yes
55.	*Clitoria ennua* (Grah)	1	1
56.	*C. ternatia* (Linn)	1	1
57.	*Crotalaria acicularis* Bush Ham ex Benth	1	1
58.	*C. albida* Heyne and Benth	N.E.	Yes
59.	*C. assamica* Benth	1	1
60.	*C. bialata* Schrank	N.E.	Yes
61.	*C. calcycina* Schrank	N.E	Yes
62.	*C. ferruginea* Grah ex Benth	1	1
63.	*C. humifusa* Grah ex Benth	1	1
64.	*C. juncea* L	N.E.	Yes
65.	*C. mysorensis* Roth	1	1
66.	*C. occulata* Benth	1	1

	Plant Name	Edible/ Non-Edible	Medicinal Use Yes/No
67.	*C. pallida* Ait	N.E.	Yes
68.	*C. spectabilis* Roth	N.E.	Yes
69.	*C. sessiliflora* (Linn)	1	1
70.	*C. tetragona* Andr	1	1
71	*C. verrucosa* Linn	1	1
72.	*Dalbergia lanceolaria* L	N.E.	Yes
73.	*D. ramosa* Roxb.	1	1
74,	*D. sisso* Roxb.DC	N.E.	Yes
75.	*D. stipulacea* Roxb.	1	1
76.	*D. thomsonii* Benth	1	1
77.	*D. velutina* Benth	1	1
78.	*D. valubilis* Roxb	N.E.	Yes
79.	*Dalhousiea bracteata* (Roxb) Grah ex Wight	N.E.	Yes
80.	*Delonix regia* (Boj ex Hook) Raf	1	1
81.	*Derris cunefolia* Benth	1	1
82.	*D. elliptica* (Wall) Benth	1	1
83.	*D. ferruginea* (Roxb.) Benth	1	1
84.	*D. robusta* (Roxb) ex DC Benth	1	1
85.	*D. scandens* (Roxb.) Benth	1	1
86.	*D. trifoliata* Lour	1	1
87.	*Desmodium alatum* DC	1	1
88.	*D. concinnum* DC	1	1
89.	*D. ferrugiinum* (L) DC	1	1
90.	*D. gangeticum* (L) DC	N.E.	Yes
91.	*D. griffithianum* Benth	N.E.	Yes

	Plant Name	Edible/ Non-Edible	Medicinal Use Yes/No
92.	*D. gyroides* (L) DC	1	1
93.	*D. heterocarpun* (L) DC	1	1
94.	*D. hetrophyllum* (Willd) DC	1	1
95.	*D. khasianum* Prain	1	1
96.	*D. laxiflorum* DC	N.E.	Yes
97.	*D. microphyllum* (Thunb) DC	N.E.	Yes
98	*D. motorium* (Houtt) Merrill	N.E.	Yes
99.	*D. oblongum* Wall ex Benth	1	1
100.	*D. podocarpum* DC	1	1
101.	*D. pulchellum* (L) Benth	N.E.	Yes
102.	*D. renifolium* (L) Schindl	1	1
103.	*D. sequax* Wall	1	1
104.	*D. spirale* DC	1	1
105.	*Desmodium triflorum* (L) DC	N.E	Yes
106.	*D. velutinum* (DC)	1	1
107 .	*D. zonatum* Miq	1	1
108.	*Dolichos biflorus* Linn	E	1
109.	*Dumasia villosa* DC	1	1
110.	*Dunbaria ferruginea* Wight and Arn	1	1
111.	*D. rotundifolia* (Lour) Merril	1	1
112.	*Entada phaseoloides* (L) Merril	N.E.	Yes
113.	*E. purasetha* DC	N.E.	Yes
114.	*Erythrina arboresscens* Roxb.	N.E.	Yes
115.	*E. fusca* Lour	1	1
116.	*E. stricta* Roxb	N.E.	Yes

	Plant Name	Edible/ Non-Edible	Medicinal Use Yes/No
117.	*E. variegata* L	N.E.	Yes
118.	*Flemengia bracteata* (Roxb) Wight	1	1
119.	*F. fruiticulosa* Wall ex Benth	1	1
120.	*F. involucrata* Wall ex Benth	1	1
121.	*F. macrophylla* (Willd) Kuntze ex Merrill	N.E.	Yes
122.	*F. semialata* Roxb.ex Ait	1	1
123.	*F. strobilifera* (L) R & Br	1	1
124.	*Indigofera artopurpurea* B& H	N.E.	Yes
125.	*I. cassioides* Rottl ex DC	E	1
126.	*I. dosua* B & H ex D. Don	1	1
127.	*I. nigerescens* King and Prain	1	1
128.	*I. spicata* Forsk	1	1
129.	*I. suffruticosa* Mill	N.E.	Yes
130.	*I. tinctoria* L	1	1
131.	*Lablab purpureus* L	E	1
132.	*Lathyrus pratensis* L	N.E.	Yes
133.	*L. odoratus* L	N.E.	Yes
134.	*L. sativus* L	E.	1
135.	*Lens culinaris* Medik	E	1
136.	*Lespedeza juncea* (L.f.) Pers	E	1
137.	*L. tomentosa* Thunb Sieb ex Maxim	1	Yes
138.	*Leucaena leucocephala* Lamk de Wit	N.E.	Yes
139.	*Lupinus perennis* L	1	1

	Plant Name	Edible/ Non-Edible	Medicinal Use Yes/No
140.	*Macro tyloma uniflorus* Lamk verdec	N.E.	Yes
141.	*Mezoneurun ennephylum* W & A	N.E	Yes
142.	*Milletia extensa* Benth ex Baker	1	1
143.	*M. fruticosa* Benth and Baker	1	1
144.	*M. pachycarpa* Benth	1	1
145.	*M. pulchera* Benth (Kurz)	1	1
146.	*Mimosa pudica* L	1	1
147.	*Mucuna bracteata* DC	E	Yes
148.	*M. cochinensis* Cheval	E.	Yes
149.	*M. deeringina* (Bort) Merril	E	Yes
150.	*M. monosperma* DC ex Wight	E	Yes
151.	*M. nigricans* (Lour) Steud	E	Yes
152.	*M. pruriens* (L) DC	E	Yes
153.	*Neptunia olearaceae* Lour	E	Yes
154.	*Oromosia assamica* Yakovlev	1	1
155.	*Parkia timorina* DC. Merrill	E	Yes
156.	*Parochetus communis* B & H exD. Don	N. E	Yes
157.	*Peltophorum petrocarpum* DC Baker ex Heyne	1	1
158.	*Piptenthus nepalensis* (Hook) D. Don	E	Yes
159.	*Pithecelobium dulce* (Roxb) Benth.	1	1
160.	*Pithecellobium gracile* Bedd.	1	1
161.	*P. heterophyllum* (Roxb.) Macb.	1	1
162.	*Priotropis cytisoides* W & A	E	Yes

	Plant Name	Edible/ Non-Edible	Medicinal Use Yes/No
163.	*Prospis cineraria* (L) Druce	E.	Yes
164.	*Prospis scapigera* L	1	1
165.	*Psophocarpus scandens* (Endl.) Verdec	E	Yes
166.	*P. tetragonolobus* (Linn) DC	1	1
167.	*Pureria phaseoloides* (Roxb)	1	1
168.	*P. thomsonnii* Benth	1	1
169.	*P. tuberosa* (Roxb ex Wild) DC	N.E.	Yes
170.	*Pycnospora lutescens* (Poir) Schindl	1	1
171.	*Rhynocosia biflora* DC	1	1
172.	*R. rothii* Benth ex Ait	1	1
173.	*R. rufescens* (Willd) DC	1	1
174.	*R. scarbaeoides* DC	1	1
175.	*Samena samen* (Jacq) Merrill	1	1
176.	*Saraca asoca* (Roxb.) de Wide	E	Yes
177.	*Sesbania bispinosa* (Jacq) W.F. Wight	1	1
178.	*S. cannabina* (Retz) Poir	E	Yes
179.	*S. grandiflora* (L) Poir	E	Yes
180.	*S. javanica* Miq.	N.E.	Yes
181.	*S. sesban* (L) Merrill	1	1
182.	*Shuteria involucrata* W& A	1	1
183.	*Smithia blanda* Wall ex Wight and Arn	E.	Yes
184.	*S. senstivia* Ait	E.	Yes
185.	*Tamarindus indica* L	E	Yes

	Plant Name	Edible/ Non-Edible	Medicinal Use Yes/No
186.	*Tephrosia candida* (Roxb) DC	N.E.	Yes
187.	*Tephrosia purpurea* (L) Pers	1	1
188.	*Uraria clarkei* Gagnap	1	1
189.	*U. crinita* (L) Desv	1	1
190.	*U. lacei* Craib	1	1
191.	*U. lagopodiodes* (Linn) Desv	N.E.	Yes
192.	*U. pulchera* Haines	1	1
193.	*U. rufescence* (DC) Schindl	1	1
194.	*Vicia faba* L'	E.	Yes
195.	*V. sativa* L	E	Yes
196.	*Vigna umbellate* (Thunb) Ohwi and Ohashi	E	Yes
197	*V. unguiculata* (L) Walp	E	Yes
198.	*V. vexillata* (L) A. Rich	E	Yes
199.	*Zorina gibbosa* Span	E	Yes

REFERENCES

Ambasta S.P. (Ed) 1986. The Useful Plants of India C.S.I.R, New Delhi.

Anon. 1976. Botanical Exploration of Tawang R.R.L. Jorhat, Assam.

Anon. 1948-76. The Wealth of India Raw Materials Vol. 1-11, C.S.I.R., New Delhi.

Arora R.K. 1980. Native Food Plants of North East Tribals. Glimpses of Indian Ethnobotany Ed. S.K. Jain, Oxford & IBH Publishing Co. New Delhi pp 91-106.

Bennet S.S.R. 1983. Ethnobotanical Studies in Sikkim. Ind. For. 109 (12): 477-481.

Bennet, S.S.R. 1983. Ethnobotanical Studies in West Sikkim. *J. Econ. Taxon. Bot.* 7(52): 317-321.

Deb, D.B. 1961. Plants of Manipur territory. Bull of Bot. Survey of India. No. 3 & 4 pp 263-350.

Deb, D.B. Flora of Tripura, Today and Tomorrow Agencies, New Delhi.

Hajra, P.K. 1981. Nature and conservation in Khasi Folk beliefs and taboos : Glimpses in Indian Ethnobotany, Oxford & IBH Publishing Co. New Delhi pp. 149-152.

Joseph, J. & Kharkongor P.A. Preliminary Ethnobotanical Survey in Khasi & Jaintia Hills pp 115-123 ibid.

Kumar, S 1995. Medicinal Plants of North East Region, Scientific Publishers (India), Jodhpur.

Kumar, S. 2004. Economic Plants of North East Region, Scientific Publishers (India), Jodhpur.

Kumar, S. 2003. Leafy and Edible Plants of North East India, Scientific Publishers (India), Jodhpur.

Nutritional Improvement of Food Legumes by breding, 1973.

Protein Advisory Group of United Nations Pulse Crops of India 1960. ICAR, New Delhi.

Sood, R.K. & Sudhir Kumar 2004. Flowers and Vegetables of India. Scientific Publishers (India), Jodhpur.

Sood, R.K. & Sudhir Kumar, 2004. Herbs : Culinary, Medicinal and Aromatic. Scientific Publishers (India), Jodhpur.

Tropical Legumes - Resources of future 1979. National Academy of Sciences (NAS) Washington, DC.

U.C.S.MPC 322